MAKING DO

Making Do

CONSERVATION ETHICS
AND
ECOLOGICAL CARE
IN
AUSTRALIA

Mardi Reardon-Smith

STANFORD UNIVERSITY PRESS
Stanford, California

Stanford University Press
Stanford, California

All photographs by the author.

Library of Congress Cataloging-in-Publication Data
Names: Reardon-Smith, Mardi author
Title: Making do : conservation ethics and ecological care in Australia / Mardi Reardon-Smith.
Description: Stanford, California : Stanford University Press, 2025. | Includes bibliographical references and index.
Identifiers: LCCN 2025017720 (print) | LCCN 2025017721 (ebook) | ISBN 9781503643635 cloth | ISBN 9781503644441 paperback | ISBN 9781503644458 ebook
Subjects: LCSH: Nature conservation—Australia—Cape York Peninsula (Qld.) | Environmental ethics—Australia—Cape York Peninsula (Qld.) | Wilderness areas—Australia—Cape York Peninsula (Qld.) | Protected areas—Australia—Cape York Peninsula (Qld.) | Human ecology—Australia—Cape York Peninsula (Qld.)
Classification: LCC QH77.A8 R44 2025 (print) | LCC QH77.A8 (ebook) | DDC 333.7209943/8—dc23/eng/20250708
LC record available at https://lccn.loc.gov/2025017720
LC ebook record available at https://lccn.loc.gov/2025017721

Cover design: Susan Zucker
Cover photograph: A storm rolling in, Hillview Station, Queensland, Australia; courtesy of the author

The authorized representative in the EU for product safety and compliance is: Mare Nostrum Group B.V. | Mauritskade 21D | 1091 GC Amsterdam | The Netherlands | Email address: gpsr@mare-nostrum.co.uk | KVK chamber of commerce number: 96249943

CONTENTS

ACKNOWLEDGMENTS

From the start, this book has been a labor of love. I went to Cape York to do fieldwork for this project, but once there, I fell so deeply in love with the people, the Country, the light, and the landscapes that my relationship to the place has transcended the intellectual work, analysis, and thought that has gone into this book. Early on in my fieldwork, a park ranger told me that once the red dirt gets in your veins, the Cape is with you for life. I was twenty-five years old when I went to Cape York, and I strongly feel that my time there re-formed me and radically changed how I think about the world, and about myself. So, first, I want to acknowledge and thank the Country on which I did this research, the lands of the Kuku Thaypan, Lama Lama, Olkola, and Guugu Yimithirr peoples. I've been held by many other places, too, in the thinking and writing that has gone into this book: Giabal and Jarowair Country, where my parents' farm is; the Country of the Gadigal people of the Eora nation, where the University of Sydney is located; and Wurundjeri Country, where I currently live and work. In all these places, Country was stolen and never ceded; I remain an uninvited guest, and I am privileged to be so.

This book project started out under the guidance and support of my mentors at the University of Sydney: Linda Connor, Åse Ottosson and Luis Angosto-Ferrandez. In particular, Linda and Åse have been with this project since its infancy. From the very first discussion about the ideas and possible field sites, to the fully realized project, they gave me their time, support, encouragement, and careful and considered critique. They helped me to push my ideas and my writing and to navigate numerous bureaucratic hurdles to be able to do the fieldwork in the way I wanted to. I am particularly grateful to them both for continuing to support me, even after their formal involvement with the University of Sydney had come to an end. I am lucky that I continue to enjoy a fruitful and supportive friendship with them both.

This book simply would not exist if it wasn't for the encouragement of my colleagues at Deakin University during my research fellowship there. Thank you to Emma Kowal for pushing me to take the plunge and allowing me the space and time to do so. With your generous mentoring, I was able to think seriously about my own research alongside our work together, and I am so grateful that you made room for this, and with such enthusiasm! Most of all, thank you to Tim Neale for guiding me through the whole process of transforming this work into a book. Thank you, Tim, for your encouragement, for your belief in me and the value of the work, and for so much tangible support that has been invaluable. I'm so grateful to have you as a mentor, a colleague, and a friend.

This book has benefited from the thoughtful feedback of so many people over the years. I must thank Veronica Strang and David Trigger, who encouraged me to take the science seriously and pay more attention to what the park rangers were doing. I was incredibly fortunate to have my book workshopped during its early stages by a group of scholars I deeply admire and respect. Thank you to Tim Neale for organizing and facilitating this, and to Jessica Cattelino, Victoria Stead, Sophie Chao, and David Trigger (again) for taking part. All the participants brought such care and insight to the work, which gave me the courage and motivation to keep pushing the project forward. The workshop was a truly transformative experience for really figuring out what the book is about, and for gaining confidence about the ideas and the writing. More recently, I have benefited from my two anonymous and insightful reviewers— thank you for your kind words and guidance for honing the ideas and the writing further. Finally, thank you to my editor Dylan Kyung-Lim White, who believed in the project and—importantly—understood how significant my photographs were to me, and to the book.

Many of the ideas in this book have been refined through the process of publication. An adapted version of the text Mardi Reardon-Smith, "Grappling with Weeds: Invasive Species and Hybrid Landscapes in Cape York Peninsula, Far Northeast Australia," *Environmental Values* 32, no. 3, pp. 249–69, Copyright © [2022] (White Horse Press) appears in chapter three (Weeds). An adapted version of the article Mardi Reardon-Smith, "Forging preferred landscapes: Burning regimes, carbon sequestration, and "natural" fire in Cape York, far north Australia," *Ethnos*, pp. 1–23, Copyright © [2023] (Informa UK Limited, trading as Taylor & Francis Group) appears in chapter five (Fire). An adapted version of the article Mardi Reardon-Smith, "The Wet: Shifting Seasons, Climate Change, and Natural Cycles in Cape York Peninsula, Queensland," *Oceania* 93, no. 3, pp. 302–20, Copyright © [2023] (The Authors. Oceania published by John Wiley and Sons Australia, Ltd. on behalf of University of Sydney.) appears in chapter six (Water).

I have received multitudes of support for this project, through each of its phases. I benefited from the Australian Government's Research Training Scheme—without this scholarship, it simply would not have been possible for me to devote years of my life to this research. The University of Sydney's Carlyle Greenwell Bequest generously funded my fieldwork, enabling me to undertake fifteen months of field research in a remote and expensive region. More recently, Deakin University supported me with a Research Development Grant that allowed me to host my book workshop.

I have been deeply fortunate to be a part of wonderful academic communities through the many iterations of this project. Thank you to my wonderful cohort at the University of Sydney and Deakin University, and to other academic colleagues and friends for endless peer (and moral) support, including but not limited to, Sophie Adams, Meherose Borthwick, Sophie Chao, Katherine Giunta, Timothy Heffernan, Benjamin Hegarty, Patrick Horne, Jaya Keaney, Jay Malouf-Grice, and Nikita Simpson.

I often think of friends as stars in the constellation that make up a meaningful life—as the people who help you to remember who you are and where you're going. I have been held, loved, and cared for by so many precious friends through the long process of bringing this work into the world. There are too many people to name but, in particular, I want to thank Imogen Gartside, Isobel Hammel, Angelica Waite, Sarah de Wit, Emika Kazama, Mashara Wachjudy, Greta Balog, Sally Molloy, Ashleigh Brooke Ralph, and a really big thank you to Sebastian Jacob-Rogers who gave me a home during my fieldwork and got to experience the highs and lows in real time! All of you have patiently listened as I've puzzled things out, have given me encouragement both gentle and firm, and many of you also visited me during my fieldwork, bringing some much-needed softness and familiarity to a big time in my life. I also want to thank Mahne Bonney and Nathan Lane, both of whom accompanied me for some of the important parts of the journey.

My family has supported me in so many ways. In particular, my parents Kate and Hugh Reardon-Smith, who very generously provided me with my trusted 1980s series Landcruiser that I drove 40,000 kilometers in during fieldwork, who have helped me move my life interstate multiple times, who let me come home to the farm to write up this research during the COVID pandemic, and who have always made becoming an anthropologist feel not only possible, but also like a good idea. Special thanks to my beautiful mum for helping me understand environmental science and proofreading an entire earlier version of this work! To the rest of my family—Susie Reardon-Smith, Han Reardon-Smith, Jill "Emmy" Kentish, Stow Kentish, Megan Kentish, Judith Kentish, Bronwyn McMahon, Bill Kentish, Khadim Mané, Ari Russell, and to our newest family member, Mustafa Mané—I love you more than I can say, and every day I'm so grateful that you are my people.

Finally, to the ringers, rangers, and Traditional Owners of Cape York, thank you for letting me into your lives and telling me your stories. I hope that in some small way this book does justice to the gift you have given me.

MAKING DO

Queensland Parks and Wildlife Service rangers
regroup after a controlled burn, 2018.

Introduction

On an oppressively hot and humid day in January I rode a quad bike along the fence line of Hillview Station, a cattle grazing property in southeast Cape York Peninsula, far northeast Australia. I was, to the best of my ability, following closely behind Justin[1]—a settler-descended grazier in his mid-forties. It had rained recently, and the new growth on the grass and trees was a bright, fresh green. The scrub[2] was reverberating with the sound of insects that had proliferated in the wet season. On the back of his bike, Justin carried tools and a long coil of wire. We stopped, periodically, to fix snapped fencing wires—dragging fallen branches and debris off the fences, picking up posts that had fallen over, and joining old fencing wires with new sections, Justin pulling the wires taut. The sky was gray and the air thick and muggy.

At points, we ducked through and under fences, crisscrossing the boundary of the station to avoid obstacles like a creek bed, a thick bit of scrub, a steeply eroded section. We came to the southwest boundary of Hillview and Justin paused. Justin was less involved in my research than his parents, and he directly addressed me about my work only rarely. Here, though, he waited for me to kill the engine on my bike before he spoke. He gestured across the boundary fence at two paddocks, separated by a further fence. "That's Aboriginal land," he points directly ahead, "and that's National Park," a wave to the left, "and we're on pastoral lease. See any difference between these three bits of land?" he asked me. I considered each paddock. They were all lightly forested savanna country, dotted with soap bush trees and termite mounds. The same species, a similar density. I shrugged. "We must all be fucking the land up the exact same way then," Justin drawled, chuckling.

We drew close to the homestead in the afternoon, dark storm clouds filling the eastern sky. The air had a charged quality, and Justin accelerated toward home. I sped up too, unsure if I would be able to find my way back to the house without him. As we rode along a flat ridge at the top of the dividing range, I looked up to see the trees ahead illuminated by the setting sun, glowing against the black of the sky. It started to rain.

This was in early 2019, and I had been in Cape York for about eight months at this point. I came to the region because I was interested in how the diversity of people in Cape York come to form, maintain, and transform their relationships with and through landscapes. Justin's mother, Pam, was one of the first people I spoke to when I began exploring the possibility of this project. I was drawn to Cape York because it had the confluence of things I was interested in: cattle, comanaged National Parks, conservation initiatives, and rich environmental and cultural heritage. Cape York is touted as a wilderness in tourism materials, and one of the last frontiers for intrepid travelers in Australia, with access limited by seasonality and unpaved roads. Trawling a series of caravanning blogs from a basement office at my university in Sydney, I came across a post about a small campground on a cattle station and some information about Pam and her family. A phone number was listed. When I called the number, Pam answered. I fumbled through my pre-prepared spiel about who I was, what I was hoping to do. I said that I was interested in cattle stations and cattle graziers. "Well," scoffed Pam. "You're about twenty years too late for that."

In some ways Pam was right. Cattle are still part of the story of Cape York, but the industry is no longer dominant in the way that it was throughout the twentieth century. Many stations have been bought by the Queensland government and transferred into National Parks and Aboriginal land. Those stations that do remain tend to be economically marginal. Yet, for people like Pam, grazing cattle remains central to how they conceive of themselves, and their role in the region. For people like Pam, grazing is still thought of as the best use of land in Cape York.

Cape York in far northeast Australia is a region of over 130,000 square kilometers (around 50,200 square miles) in size that is renowned for its environmental and cultural values. These values have been recognized both nationally and internationally, with the Quinkan rock art galleries of the sandstone escarpments in central Cape York achieving Australian Heritage listing and the tropical savannas, coastal aeolian dunes, and aquatic and rainforest ecosystems of the region being considered for World Heritage listing (Chester 2010; Australian Government 2012; Hitchcock et al. 2013). Sparsely populated, geographically remote, and accessible to nonresidents only during the annual dry season, the region is frequently touted as one of Australia's few remaining wildernesses. It is home to a range of endemic and endangered species, visually stunning savannas and plains, forests, sandstone escarpments, the so-called wild rivers, and ecologically significant wetlands and coastal estuaries.

Like other biodiverse regions in the world, Cape York sits at the nexus of important political struggles as well as social, cultural, and environmental changes. Cape York's significance as a series of protected bioregions of national concern cannot be underestimated, nor can its importance as a region rich in a diversity of Aboriginal cultural heritage, contemporary lifeways, and Aboriginal political struggles. The region—which has a high proportion of Aboriginal people in comparison to the general population of Australia—has been a testing ground for new forms of land tenure, for environmental protection legislation, and for Aboriginal economic development projects. While throughout the twentieth century the cattle grazing industry dominated the region and most land was held under pastoral lease, the situation differs substantially today. Much of Cape York is now recognized through some form of Indigenous land rights. For Aboriginal peoples who have sustained struggles to have their rights over land recognized for decades, these changes are understood as gradual and the result of incredible amounts of labor and care—frequently undertaken by older people who have since passed away. For White cattle graziers, these changes are framed as rapid, occurring with little foresight, and are read broadly as an attempt to "lock up" the region from use. While their numbers are dwindling, graziers remain an important part of the social and economic fabric of the region. Care for land occurs across different tenure types. Pigs, cattle, weeds, fires, and waterways are not constrained by land tenure boundaries, and thus different groups of people are brought together in caring for, managing, and seeking to control aspects of the land. Importantly, the management agreements that sit across tenures and the permeability of boundaries between them result in an extension of the reach of the state government (often in the form of the Queensland Parks and Wildlife Service [QPWS]) into both Aboriginal and settler-descended peoples' lives.

This book explores the coproduction of care between a variety of people, forms of care directed toward place, animals, plants, elements, and ancestors. In the intercultural contexts of the multiethnic cattle grazing industry and comanagement of protected areas, environmental knowledges and ways of practically relating to land in Cape York emerge and are continually transformed in iterative interactions between different aggregates of humans. Throughout this book, I investigate how people use their environmental knowledges in an effort to control landscapes, species, and other people, with particular attention to the asymmetrical relations that shape how people care for land, seek to control it, and remain complicit and culpable in environmental harms.

These confluences of different forms of land care and knowing matter in thinking through the legacies and possible futures of landscapes shaped by settler-colonialism through the dispossession of Indigenous peoples and interruption to Indigenous land management practices, the development of agro-industry, and the spread of introduced species. Cape York is a place where diverse groups of people have deep and enduring connections to land, based on a mixture of ancestral connection, careful observation, and laboring on the land. Yet Cape York is also a place that has national and global significance as the site of multiple protected bioregions, with land held as national parks, nature refuges, marine parks, and under Indigenous Land Use Agreements. Despite the very human role in shaping the landscape of Cape York (through fire, through the distribution of species, through care and dwelling) (Langton 2002), the region remains widely thought of as a "wilderness." Cape York is home to multiple famed national parks comprising a land area of 43,000 square kilometers (around 16,600 square miles), as well as privately run nature conservancies and conservation stations. In this context, what counts as "natural," "native," and "belonging" matters (Trigger 2008; 2013)—as does who gets to decide how species and people are categorized and, accordingly, how they are controlled.

I am concerned with complicating commonsense understandings of "wilderness" places and spaces. Throughout the book, I make three related claims: that environmental knowledge and ethics are coproduced relationally; that conservation and land-managing actors do not conform to management plans and broader visions for the region as a wilderness space; and that introduced species proliferate in unexpected ways, bringing new relationships into being that require new kinds of relationality such as killing as a form of care. I'm interested in the unintended consequences and collaborations of interspecies and intercultural mingling, considering what is inherited and left behind, and what is imagined for the future. What will become clear in this book is that the presumed categories through which we might understand a place like Cape York do not play out as expected on the ground. Environmental knowledges and ethics, human-environment relations, and interpersonal human relationships emerge through practice—through labor and care—in ways that are often surprising, sometimes confounding. Park rangers who seek to protect and conserve environmental values wage an ongoing battle against invasive pigs, while also seeking to censure, control, and restrict the actions of pig hunters in the park; a self-professed antigreen grazier devotes decades to recording detailed observations of an endangered parrot, even though she is aware her cattle present the most significant risk to the bird; Aboriginal Traditional Owners embrace imperfect joint management arrangements, syncretizing different knowledge systems with a pragmatism that demonstrates an extraordinary capacity to adapt to new circumstances. In these instances, and others, the people who labor to make livable lives and cultivate workable landscapes in Cape York endure, persist, and make do through effort, care, and practice.

Broadly speaking, the book asks what happens when shifting aggregates of humans and nonhumans exist in the world in unexpected ways, complicating commonsense ideas about regions widely considered to be wildernesses. I ask why conservation initiatives and regimes falter and fall apart and investigate how complicity and care can coexist in human-environment relationships. What happens when invasive species proliferate? When park rangers do what they want? When graziers let their cattle through broken fences? What happens to land management when the land and its inhabitants refuse to be managed? When people, plants, and animals refuse to sit still? Even those people explicitly employed to do conservation work have complex and contingent relationships—to land, to people, to nonhuman species.

Increasing amounts of land around the world are being set aside for protection and conservation, and yet conservation outcomes continue to remain poor. Biodiversity continues to be lost, vulnerable species continue to slide into extinction, and invasive species continue to proliferate and cause damage to sensitive ecosystems. Conservation projects frequently fall short and have impacts on local peoples that ripple out (Milne 2022; West 2006; Doolittle 2005). At their worst, conservation projects lead to people being excluded from protected areas and decision-making, having their resource bases dramatically restricted, and having their relationships to land transformed.

There was a time when I thought this book was about how to do conservation *better*. It's not. Instead, this book is about care for land, and care among humans, and between humans and the more-than-human, that can be understood as "otherwise" (Povinelli 2011; Povinelli 2016). It is about care for land that presents a counterpoint to conservation as the dominant shape through which to conceptualize good land management, and good human-environment relations. Care as an otherwise to conservation takes us outside of the politics of purity that so constrain conservation efforts and opens up space for human-environment relationships that are messy, tangled up, and imperfect. Human-environment relationships in which complicity in environmental harms sits alongside (and sometimes functions to compel) care for land. In which forms of care can be generative for a multitude of others *and at the same time* be self-interested. This book is asking how is it that we know each other, and how do we know the places in which we live, and how do we care for them, among the ruptures of colonization and planetary failure.

A central through line in the book is the idea of "making do." In a sense, making do is a distinctly rural sensibility, related to an ethic of enduring and persisting in among trying conditions and uncertain futures. As will become apparent, the people who populate the pages of this book are not laboring toward a horizon of deliverance; there is no sense that a utopic future is around the corner, or that things will necessarily *get better*. Instead, people labor to make do in an environment that is tough to live in, a place in which people must endure climate extremes, the proliferation of weeds and pests, unpaved and frequently impassable roads, the challenges of remote area living, and life at the economic margins. As I hope to show in this book, people make do in this challenging place through their relationships of care.

In this place, making do is entwined with the caring labor that goes into making landscapes workable and lives livable. Even though many of the people that this book is about work in jobs that are, technically, conservation and land management roles, I argue that their caring labor frequently sits alongside and outside of the formal structures and internal logics of conservation. Sprawling, messy, sometimes violent and often self-interested, these forms of care—which I gloss as making do—are what hold together the humans, nonhumans, and landscapes of Cape York.

On Interculturality

Cape York has been a site of significant anthropological research, particularly throughout the twentieth century. In Australia, much anthropological scholarship of this era focused squarely on the lives and sociopolitical conditions of Aboriginal people. In Cape York, anthropologists have provided rich and important accounts of the distinctiveness of Aboriginal lifeworlds, cultural practices, and political struggles (e.g., Sutton 1978; Von Sturmer 1978; Chase 1980; Anderson 1985; Martin 1993). However, Aboriginal people in Cape York live and work alongside a range of non-Aboriginal people. In my time in Cape York, I found that there are important commonalities among Cape York land managers: they all must interact with the nonhuman elements, such as cattle, weeds, fires, and monsoons that are active forces in the landscape. I seek to extend our understanding beyond presumed sociocultural boundaries by attending to the relational and intercultural processes through which Cape York residents come to value, care for, and secure a livelihood in the changing environment. Significantly, I find that the values, priorities, and practices of these groupings are not necessarily oppositional in an Aboriginal–non-Aboriginal binary, but they converge and diverge to varying extents depending on structural conditions, the particular social and physical context, and the personalities involved.

I find Francesca Merlan's (1998; 2005) "intercultural" approach to contemporary Aboriginal peoples' lives to be particularly productive for thinking through the Cape York context. This "intercultural" approach has emerged as a way to theorize the situation of "difference-yet-relatedness" that is reproduced in a settler-colonial state like Australia (Hinkson and Smith 2005, 157). Rather than assuming that Aboriginal people and non-Aboriginal people come to encounters as already pre-formed, culturally different, and bounded entities, the intercultural approach understands sociocultural difference and forms of identifying as fundamentally relational, as emerging in social relations, iterations, and practice (Ottosson 2010). To borrow Donna Haraway's (2008, 25) phrase, it is through "the dance of relating" that cultural ideas, norms, and practices are reproduced or transformed. Everyday social interactions can operate as a zone of reproduction and change, and it is in this zone that culture is worked over, reproduced, and transformed (Merlan 2005, 169–70).

In her review of contemporary Australian Aboriginal anthropology, Tess Lea (2012) notes anthropology's widespread use of Merlan's concept of the "intercultural" but highlights how the institutions of White Australia frequently remain obscure in ethnographic accounts, broadly painted as "the state." As she writes, "too often, only one side of the colonial relation is intercultural, despite decades of postcolonial critique" (Lea 2012, 191). In this book, I seek to pay careful attention to the lives and experiences of settler-descended cattle graziers, park rangers, and Aboriginal Traditional Owners alike. The histories and possible futures of Cape York are entwined with each of these groupings of people. Importantly, Merlan's (1998) intercultural approach does not foreclose the possibility of attending to asymmetrical power relations. People do not necessarily come to interactions on equal footing. For instance, Western scientific knowledge continues to be dominant in land management planning and practices, even in the context of formal joint management. As has been described in a variety of other contexts, Aboriginal environmental knowledges continue to be treated as requiring validation through the Western scientific process to be deemed "knowledge" (Carroll 2015; Hutchings 2014; Liboiron 2021; Smith 1999, 66).

Cape York is a rich site of interaction, with people pulled—at times forced—into engagement through the historically multiethnic pastoral industry, contemporary land tenure changes, management agreements and regionwide pest control and fire management. In these spaces of sometimes uncomfortable modes of relating, environmental knowledges and values have been and continue to be coproduced, but in ways constrained and shaped by unequal power relationships.

But relationships between humans are only part of the story. Plants, animals, ecosystems, and elements shape social worlds, too.

On Care

It was about halfway through my fieldwork, on a late afternoon in early December 2018, that I had one of my most personally difficult experiences in Cape York. I was driving across a flat, sandy plain, lightly forested with shrubs, small trees, pandanus, and grasses. My dissertation supervisor was in the car with me. We were on our way north to the outstation community of Yintjingga, where we planned to stay the night, sleeping bags rolled out in the Community Building—as I'd done several times before. We had left the small town of Mount Molloy relatively early, but the drive is long, and we had taken several breaks, so we were driving through Lama Lama Country at sunset. I hated driving at this time, especially on a road like this one, which experienced little traffic. Wallabies come out to feed toward the end of the day.

I didn't see the wallaby that my car hit—or, more accurately, that hit my car. I heard it, though. It hit the rear left tire with such force that it bent the wheel guard. I had to enlist the help of a friend's brother to remove the bent piece of metal later that day when we made it to Yintjingga. I pulled over and stepped out of the car. I could see that the wallaby was still alive, but that it wouldn't survive for long. We were hours' drive away from a veterinarian, and I knew that they would only euthanize the wallaby even if we could transport it there. Panicking but with a grim realization of my responsibility, I retrieved a hammer from the back of my car. It was the only tool I could think to use. I walked over to the frightened wallaby. It tried to scramble away but with its broken bones it could move very little. What follows is seared into my memory. I held the wallaby steady. I spoke to it gently, while it writhed and foamed at the mouth. I hit its skull with the hammer. I was nervous, and the first blow wasn't hard enough. I don't know how many times I hit the wallaby. A few, I guess. I kept my hand on its tiny ribcage, feeling its heartbeat through soft fur and narrow, jutting bones. I waited until there was no heartbeat and returned to the car, shaken, crying. The rest of the drive to Yintjingga I could hear the bent wheel guard thumping on the tire. I had visions of the rubber shredding, but we arrived at the community, tire intact. Driving south a few days later, I kept an eye out for the wallaby's body. The body was gone, returned to the nutrient cycle.

I killed a lot of animals while in Cape York, and watched others being killed also. Mostly my killing was accidental and from a distance. I hit wallabies in my car and, once, a small bird. I mowed over a snake. I caught a fish (only one, I never really got the hang of fishing despite significant tutelage) and let the women I was with deal with its death. I stood on a few cane toads but I'm pretty sure that they survived. But my encounter with this wallaby was the only active killing that I was a part of. It was the only situation where I made the choice to act, the choice to kill. This killing was a kindness, but my sense memories of the incident are strong enough to put me on edge even now. I find it difficult to drive along a rural road at sunset without flashes of my hand on the wallaby's ribcage (its warmth, the feeling of its beating heart) passing through my mind. In this book I think a lot about how relations of care can involve violence, responsibility, and difficult, less than satisfying, decisions.

I conceive of care as existing in multiple registers, scales, and forms. Across Aboriginal Australia, the concept of "caring for Country" is a through line in land rights and land management endeavors (Kingsley et al. 2009; Altman 2012; Langton 2002). As Noah Pleshet writes, "'caring for Country' has emerged as a coherent elaboration of a longstanding trope" that makes Indigenous relationships to and management of land legible to the state and the broader public alike (2018, 184). The phrase functions as a kind of shorthand to describe the obligations and reciprocities that characterize Aboriginal relationships to land. "Caring for Country" is simultaneously an ethic, an ontology, a philosophy, and a complex of material practices (Graham 1999; Graham and Maloney

2019; Watson 2009, 2018). Kombumerri and Wakka Wakka philosopher Mary Graham writes that, for Aboriginal people, "custodianship is thus a philosophy, not just a green solution to environmental degradation . . . essential to this system is the fact that Aboriginal personal identity extends directly into land itself; this helps to explain why knowledgeable members of the Aboriginal community continue to assert that, *'the land is the Law'*" (1999, 116, emphasis in original).

"Caring for Country" involves care as a multiplicity of things: care for biodiversity, ecosystems and specific species, care for ancestral spirits who dwell in the landscape, care for culture and bodies of knowledge, care for living Aboriginal people, and care for future generations of Aboriginal people, too. As one Aboriginal traditional owner told me, "respect for the land is actually respect for the old people [ancestral spirits]" who dwell there. Caring for Country requires people to both care about and do the material labor of caring, through firing the landscape, walking, dwelling, speaking to ancestors, fencing, monitoring, transmitting knowledge to young people, and dealing with invasive pest species, all of which are activities that work to link people with the land (Yibarbuk et al. 2001; Rose 1992; Povinelli 2016). The idiom of caring for Country is invoked in support of land-back initiatives, lending discursive weight to the argument that Aboriginal Traditional Owners are the people best placed to work on and with the land (Pleshet 2018). As Pleshet (2018) and others (Altman 2012; Rose 1996) have noted, the notion of "caring for Country" has become politically salient for Aboriginal people in Australia. My interest in this idea, though, is how "care" links the labor of Aboriginal people to other people engaging in environmental and land management.

In Cape York, as in many places, there are a diverse set of people involved in land care work. Aboriginal people care for Country alongside a swathe of others, each of whom do caring work in various ways. While caring for Country often involves loving Country, it is vital to remember that to care is different from to love. In her important work, which brings care to bear on multispecies relations, Puig de la Bellacasa builds on Joan Tronto's influential definition to argue that care is "everything that we do to maintain, continue and repair 'our world'" (2017, 3). Puig de la Bellacasa suggests that the notion of care encompasses both the capacity to care about (that is, an affective and ethical position that is akin to worrying for) and to care for in a material sense, by doing the sometimes tedious maintenance work that relations of care require. In this, Puig de la Bellacasa (2017, 4–5) is emphasizing that while "to care" can be a moral stance, "caring" itself demands much more than this. Care is a labor practice. It is enacted through labor, through work, through doing.

I define caring labor quite broadly in this book. It is the cumulation of experiential and embodied knowledge that is put to work in firing landscapes, in breeding cattle, in fixing fences, in controlling invasive species, in monitoring rainfall, in making a living on the land. While often economically productive, this kind of labor differs from that which is sold in waged work, because receiving an income is not the sole goal of this kind of labor. As I have written about elsewhere, labor and hard work is valued in Cape York because of its relationship to land, to the mixing of sweat with the soil (Garbutt 2011; Geschiere 2009; Reardon-Smith 2023b)—a value that is held in common by White cattle graziers and Aboriginal Traditional Owners. Following Kathi Weeks (2011), I use *work* and *labor* interchangeably in this book, although my reasons for doing this are slightly different. I do this in order to emphasize how these distinctions collapse into each other in a context like Cape York, where working and laboring are, all at once, about securing a livelihood, connecting to place, persisting with an economically marginal yet valued way of life, and enacting care for land, nonhuman species, and ancestors.

It may strike some readers as strange that gender is not the lens through which I analyze caring labor. My understanding of both care and labor owe much to feminist analysis, and yet in this story, and in this particular place, attending to gender is not the most apt intervention. In part, this is because the work that people do does not always occur along gendered lines; husbands and wives run cattle together, and men and women work alongside each other in the ranger teams. This is not to say that gender does not structure how people relate to work, or to place, to each other, or to the idea of expertise and authority, but it is not the frame I have taken in this project. Instead, I have chosen to focus on how economic relationships to land and proximity to formal conservation projects work to shape how people care for land, ecosystems, species, and each other.

For the people in this book, caring labor is not just about maintaining, continuing, and repairing worlds, but is also about bringing into being imagined futures. Caring labor spans temporality; it is simultaneously about reckoning the with the past, enduring the present, and fostering workable landscapes and livable futures. These futures are not necessarily imagined as looking particularly different to the present day, but a concern with legacy is held in common across differently socially located people, nonetheless. People enact caring labor to make do, to make livable, and to make workable. In considering what they frame as radical care, Hobart and Kneese write that "mobilizations of care allow us to envision what Elizabeth Povinelli describes as otherwise" (2020, 3). That is, care is about envisioning imagined possible futures and bringing these possible futures into being.

Importantly, Puig de la Bellacasa reminds us that care is not straightforward and is not necessarily synonymous with warm and loving affective relations. Sometimes to care is difficult. Sometimes it requires uncomfortable and unsatisfying decisions to be made (van Dooren 2014). And sometimes care can necessitate forms of violence (van Dooren 2014; Bocci 2017). As Martin, Myers, and Viseu (2015, 627) point out, caring is not innocent. To care is always to be involved in complexity and often to be drawn into relations that are asymmetrical.

In Cape York, caring for ecosystems, landscapes, and biodiversity rubs up against the presence of invasive species, and throughout the book I explore what is demanded or required of land managers. I think about how killing is a kind of care labor. People kill pigs, weeds, and other pest species in the hope of protecting waterways and wetlands, among other things. Naisargi Davé (2017) speaks about the role of the witness in human-animal interactions; the ethical imperative (on the part of the human) to not avert their eyes from the suffering of nonhuman others. I seek to consider the suffering nonhuman as not just a singular animal, or species, necessarily, but an entire ecosystem. Throughout this book I consider what responsibilities and response-abilities (Haraway 2008) are being asked of land managers in their work when they are confronted by landscapes damaged by overgrazing, improper burning regimes, the spread of weeds, the digging of pigs. Care involves "a willingness to respond" (Martin, Myers, and Viseu 2015, 634) and, despite the diversity among land managers in Cape York, a willingness to respond to the landscape itself is held in common by all. The proliferation of invasive species in Cape York demands response from those tasked with caring for the land there, and this particular situation and dynamic makes the need and willingness to respond clear. While holding in mind the danger of temporal thresholds, of seeking to return landscapes to a pure "natural" nativeness (Shotwell 2016), there is a requirement—a demand—to not turn away from the very real damage that invasive species (in particular, pigs, cattle, and weeds) do to the sensitive ecosystems and endemic species of the region.

On Complicity

Throughout this book I grapple with a desire to trouble and complicate the assumption that native equals good and introduced equals bad, and yet simultaneously hold in mind the very real damage that certain species, brought to the region not of their own volition, do to the ecosystems and to the possible futures of Cape York. Land managers today dwell in inherited landscapes. As Thom van Dooren reminds us, "care always thrusts us into an encounter with ghosts, our own *and* others" (2019, 90, emphasis in original). The notion of inheritance is, as van Dooren asserts, productive for both the biological and social sciences—particularly when brought to bear on threatened ecosystems or species with uncertain futures. He suggests that "the movements of genes, ideas, practices, and words between and among generations are all tangled up with one another, unable to be isolated into separate processes or channels of inheritance" (2019, 88). In light of this, we can't think of colonization, conservation biology, the nature-culture binary, land rights, national parks, intercultural relationships, erosion, cattle, introduced species, burning regimes, climate change, and endangered parrots as separate and discrete things. They are all knotted with each other, shaped by a common context.

Thinking about and with complicity, both as a counterpoint to care and as a driving force compelling people to engage in acts of care, is a central tension explored in this book. In his work on farmers grappling with aquifer depletion in America's Midwest, Lucas Bessire (2021) explores the tension between inheritance and culpability. Tracing his own family's role in the current ecological crisis of the Plains, Bessire proclaims that even he is complicit, even he has played a role in the depletion of groundwater resources by virtue of being born to a farming family. While acknowledging that he did not directly carry out the practices that have left a lasting imprint on the landscape and its waters ("I did not pump any water or shoot any buffalo or spray any toxins" [2021, 175]), Bessire suggests that inheriting this legacy engenders responsibility. As will become clear throughout the chapters of this book, the current land managers of Cape York have inherited a landscape already scarred—for some, by their own predecessors; for others, this scarring occurred as part of the same colonial project that dispossessed, forcibly removed, enslaved, and murdered their ancestors. Complicity and culpability are not inherited equally.

Indeed, some land managers today remain complicit in ongoing environmental harms in some ways. Does this diminish their ability to care? Does this undo any benefits of the caring work that they do? Introduced species, too, are shaped by their inheritance. Whether they were brought to the region as "four-legged soldiers in the army of conquest" (Rose 2004, 86) (i.e., cattle) or have trickled beyond the confines of their imagined existence (i.e., pigs, weeds), these species and the humans who seek to control them are entwined with the histories of the region—of colonialism, of imaginaries of the north. These inheritances are important, as are the ways in which ongoing relations of care can bring possible futures into being (Puig de la Bellacasa 2017, 83). Puig de la Bellacasa asks, "for what worlds is care being done for?" (2017, 64–65). Care can pull apart, remake, and transform relations. Like all forms of relationality, care is enacted in the doing.

Throughout this book, the reader is introduced to a host of human and more-than-human characters who coalesce in various ways. While I follow Galvin (2018) in investigating what it is that these more-than-human beings "do" in the social worlds they interact with and inhabit, I turn back, again and again, to the human role in all of this. I seek to hold onto the muddiness and messiness of the human part of human-environment relationships, and dig into the uneasy ways in which care, complicity, and conservation coexist. An important concept that I develop in this book, and particularly in chapter two, is that of "workable landscapes." Somewhere between a working landscape and a livable nature (Hamilton 2018; Scaramelli 2021), I conceive of workable landscapes as places in which people can make their lives and livelihoods while at the same time laboring to care for land, ecosystems, and species. As with working landscapes, I seek to emphasize the significance of economic relationships to land, even in the context of landscapes that are protected and do not play host to agro-industry. There is something here akin to the concept of Country in Aboriginal Australia; land that is socialized and humanized, relied on for survival, and cared for by its custodians. In Country, and in workable landscapes, the presence of people to do the caring labor of looking after environments is paramount. "Workable" also gestures toward the imperfect, and the incomplete. The good enough. The making do. The necessary compromises, the mingling of care and complicity, that come with trying to make a life in a still-remote, economically marginal place. These compromises entail the kind of contradictions that people can live with. As Davé writes, comfortability with these contradictions means that people are "not *all cramped up* with a concern for boundaries" and, accordingly, that they "are not at war with the world" (Davé 2023). To make workable landscapes, and to make do, means to enact care for land that is more pragmatic than pure.

An Introduction to Southeast Cape York

The inland of Cape York is savanna country with rocky ridges and plateaus forested with towering stands of Cooktown ironwoods, as well as smaller scrubby plants. Farther toward the coast, the landscape flattens and is composed of more densely forested river systems, salt pans, swamps, and estuaries. Throughout the region are striking termite mounds, dotting open plains and looming like thousands of gray and terra-cotta gravestones. During the dry season, the colors are muted—sun-dried golden grasses, silvery-leaved shrubs, and glittering sandy soils. In the wet season the landscape is transformed, with swollen rivers and swamps and iridescent green growth.

Cape York is tropical, with a distinct wet and dry season. The dry season normally lasts from around June to November, and the wet season from around December to May. The average annual temperature for Cape York is 26 degrees Celsius (78.8 degrees Fahrenheit), varying only a few degrees throughout the year, while the level of humidity varies significantly. The region's annual average rainfall is 1,305 millimeters (51.4 inches), with the vast majority of rain falling between October and March during the monsoon (Department of Environment and Science 2019). What this feels like is a hot and pleasant dry season, and an oppressively hot and steamy wet season.

Aboriginal people have inhabited Cape York for an estimated 37,000 years, and many Aboriginal people continue to have strong relationships to their ancestral homelands (Neale 2017). European settlers first came to the region in the late nineteenth century, initially to mine gold and, later, to take up pastoral leases and run cattle (May 1994; Cole 2004; Loos 1982). Cape York Peninsula is widely recognized as one of Australia's most violent and protracted colonial frontiers. As in other places, such as America's Midwest (Bessire 2021) and parts of Central and South America (Grandia 2012), the introduction of agro-industry transformed the lives of indigenous peoples, excluding people from their land and curtailing their ability to access resources (Queensland Land Tribunal 1996). The grazing industry remained dominant through most of the twentieth century, and many Aboriginal people who were displaced during the overt frontier violence that came with settler incursions were employed (albeit for no or low wages) on cattle stations as stock workers or domestic servants. Many Cape York graziers will admit that the region is good "breeding country," but not good "growing country," and it is difficult for people to make a living solely from a cattle business. For many years, the profitability of cattle stations in the region relied on an un- or underpaid Aboriginal workforce that was akin to indentured labor. Given the racist government restrictions under which Aboriginal people lived at the time, Aboriginal people had little freedom to leave the cattle stations where they were contracted to work, regardless of the conditions (May 1994). The introduction of Equal Wages legislation in the late 1960s and construction of fencing on cattle stations resulted in a significant decrease in Aboriginal employment in the cattle industry, as graziers either could or would not pay their Aboriginal employees the award wage, resulting in many Aboriginal people becoming reliant on government assistance. Since then, the number of profitable cattle stations in the region has been in steady decline. Even with the introduction of management tools like fences, nutritional supplements, and helicopter mustering, small-scale farming cattle on Cape York is barely viable without an unpaid Aboriginal workforce.

Australia's beef industry has generated its fair share of controversy, both in terms of perceived animal rights violations and environmental impacts. In Cape York, such controversy is made acute by the region's geographical isolation, economic marginality, and proximity to the World Heritage Listed Great Barrier Reef. Cattle in northern Australia are generally pasture-raised, meaning that cattle spend most of their lives as relatively free-ranging beasts. As such, these animals exist in a vastly different context to, for instance, the industrial hog farms of the United States, which Blanchette (2020) has written about in horrifying detail. However, many of the practices and processes that graziers engage in come under public scrutiny. Animal rights advocates have taken issue with the practice of "tipping" cattle (trimming their horns) and with livestock transport, in particular live export (which I detail in chapter one) in which cattle are shipped overseas to be slaughtered.

While graziers frequently express fears about animal rights activists, who they perceive as bent on destroying their livelihoods, perhaps more significant—for the public and in terms of the actual forces that shape graziers' lives—are the environmental impacts of cattle raising. Many of these are generalized and relate to any place where cattle are farmed; impacts such as methane emissions, soil erosion, nutrient depletion, and deforestation (Sakadevan and Nguyen 2017; Pierrehumbert and Eshel 2015). However, specific to north Queensland are the impacts of soil compaction and sediment runoff into the ocean, which threatens the well-being of the Great Barrier Reef. The cattle industry is positioned by a wide range of actors, including some Queensland Parks personnel and environmental activists, as damaging for the reef. There has been significant scientific research done on the impacts of cattle farming for the reef and, conversely, the positive impacts of de-stocking stations in terms of sediment runoff[3] (Bartley et al. 2014; Koci et al. 2020; Brodie et al. 2012). Coral reefs are hugely significant marine ecosystems, often referred to as the "rainforests of the ocean" (Braverman 2018). They are incredibly biodiverse, supporting an estimated 25 percent of all marine species. In addition, coral reefs, in Australia and elsewhere, function for the scientific community and the public more broadly as a "canary in the coal mine" in relation to climate change impacts (Braverman 2018). Given the known impacts of cattle grazing on the health of the already-struggling Great Barrier Reef, cattle farming in the Cape York region is hugely controversial. This is the case even as most runoff in the Cape flows westward, away from the reef, and the most significant catchment in terms of reef impacts is the Burdekin River basin region, located to the south of Cape York. As an imagined site of "wilderness," biodiverse and primordial, Cape York is a region in which industries that are perceived to be environmentally destructive attract significant scrutiny. It is a region where successive state governments have sought to secure votes through making promises to conserve and protect wild, biodiverse tracts of land from the polluting influence of people (Holmes 2011b).

Since the 1970s, the Queensland government has declared a number of national parks and protected areas in the region (Rigsby 1981; Holmes 2011a; 2011b). The creation of one of the earliest parks in the Cape—Oyala Thumotang National Park, formerly Archer Bend National Park—was fraught. In the early 1970s, Wik-Mungkan man John Koowarta attempted to purchase the pastoral lease over a section of his traditional homelands in western Cape York from the Archer River Pastoral Holding (Bennet and Sheehan 2021; Department of Environment and Science 2021a). The American businessman who held the lease agreed to the sale, which would see the land returned to Mr. Koowarta and several other stock workers who were Traditional Owners for the region, supported by the Aboriginal Land Funds Commission. In a decision that was later found by the Supreme Court and the Human Rights Commission to be unlawful and an example of racial discrimination, the sale of the lease was blocked by the conservative Bjelke-Peterson Queensland Government (Bennet and Sheehan 2021). The reason provided was that "the Queensland Government does not view favorably proposals to acquire large areas of additional freehold or leasehold land for development by Aborigines or Aboriginal groups in isolation" (Koowarta v. Bjelke-Petersen 1982). While the case proved to be an important early test for the 1975 Racial Discrimination Act, the ruling occurred too late for Mr. Koowarta to gain title over his land. The Bjelke-Peterson government declared the area of the pastoral lease as Archer Bend National Park in 1977, effectively removing the possibility of the land being returned to the Traditional Owners (Department of Environment and Science 2021a).

With its declaration working to forcibly exclude Aboriginal people from gaining control over their land, the contested creation of this early national park marked the beginning of tensions between Aboriginal people and the environmental movement in far north Queensland, tensions that came to national prominence during the Daintree road dispute (Anderson 1989) and "wild rivers" controversy (Neale 2017).

National parks in Australia follow the preservationist "Yellowstone model" of conservation, resulting in protected areas in which local people are excluded and their access to resources severely constrained. According to Descola (2013) and Cronon (1996), the romanticism and reverence toward a separate "Nature" that underpins preservationist conservation emerged in the West as a response to large-scale industrialization. As Descola and Pálsson (1996) argue, the ontological category of "nature," as separate from culture, is unique to Western societies. They argue that despite notions of "wildness" existing in some form in many societies, what is significant to a Western ontology and epistemology is the reliance on a binary framework: nature in opposition to culture. As various Indigenous scholars from settler-colonial contexts have discussed (Graham 1999; Carroll 2015; Liboiron 2021), Western ontology and philosophy tends to frame land as a resource, rather than a set of relations. A framing of land-as-resource, thus, situates humans as polluting and threatening. These ontological assumptions underpinned the preservationist model that national parks are based on. Among the American naturalists who formed the Sierra Club and were key to the development of the national park model was a sense that "Nature" was external, sublime, to be enjoyed at a distance, and—importantly—under threat (Tsing 2005, 95–96).

Since their inception, national parks around the world have instigated encounters between conservationists and Indigenous peoples while simultaneously extending the reach and power of the project of settler-colonialism (Carroll 2014; Jacoby 2014; Spence 1999). The first national park in the world to be declared was Yellowstone in the United States, and its designation necessitated the removal of the indigenous Shoshone, Lakota, Crow, Blackfoot, Flathead, Bannok, and Nez Percé peoples from their land (Nabokov and Loendorf 2004; MacDonald 2018). The forced removal and exclusion of Indigenous peoples from protected areas served to create the fiction of an untouched, pristine, and wild Nature ready to be enjoyed, at a distance, by cosmopolitan visitors (Jacoby 2014; Spence 1999). This practice of dispossessing Indigenous peoples for the purposes of conserving landscapes is evident across the world in a variety of contexts (Tsing 2005; Doolittle 2005; West, Igoe, and Brockington 2006; West 2006).

The creation of the early national parks in Cape York relied on this narrative of parks needing to be protected from the polluting influence of humans but was also driven by an explicit resistance by the Queensland government of the time toward Aboriginal land ownership in general. It is somewhat ironic that the deeply right-wing Bjelke-Peterson government invested in conservation in Cape York to this extent, even if the impetus was embedded in racism.

With global shifts toward considering the rights of Indigenous peoples, the preservationist approach to conservation has been critiqued by Indigenous activists and scholars alike, resulting in a move toward including Indigenous peoples in the management of protected areas. As a result, recent decades have seen the establishment of formal co-management arrangements in a number of places around the world, including Canada (Nadasdy 2003) and Malaysia (Doolittle 2005) among others. While joint management has existed in Australia's Northern Territory since the 1970s (in Kakadu National Park) (Haynes 2009), the first jointly managed park in Cape York was declared only in 2008. As such, joint management is still a new reality for the region, and many of the difficulties that emerge when different knowledge systems and land management practices come into conversation on unequal footing are present here.[4] Emerging out of the land rights movement and struggle for recognition (Coulthard 2014), joint management remains a site of contestation and negotiation, as Indigenous groups struggle to enact care for land within a highly bureaucratized system of environmental governance (Carroll 2015; Nadasdy 2003).

In 2010, around half of the area of Archer Bend National Park was revoked and in 2012 was returned to the Oyala Thumotang Land Trust, and the same year the park itself was renamed Oyala Thumotang National Park and ownership transferred to the Wik-Mungkan, Ayapathu, and Southern Kaanju peoples (Department of Environment and Science 2021a). At the hand-back ceremony, the premier of Queensland at the time made an apology to the Traditional Owners. He is quoted as saying that "thirty-five years ago a great injustice was perpetrated. Today we put that right. So again, my apologies to those who have suffered"

(Caruana 2012). Having died in 1991, Mr. Koowarta did not live to see his land returned to Aboriginal ownership (Bennet and Sheehan 2021).

Cape York is a site of conflict over appropriate uses for land, proper land management practices, and valued environmental knowledges. Aboriginal Traditional Owners, settler-descended cattle graziers, and the Queensland Parks and Wildlife Service (hereafter Queensland Parks) are each involved in the land management industry. While much environmental knowledge is held in common and many land management practices shared, groupings of people tend to converge and diverge depending on the setting and varying across time. Importantly, difference does not necessarily occur along racial lines. Relations between people are frequently asymmetrical, and in managing and caring for land, the differently powerful positions that people, groups of people, and institutions occupy become apparent. Queensland Parks is the second largest law enforcement organization in Queensland, after the Queensland Police Force, and ensuring compliance is a significant part of Queensland Parks' role in the region, especially considering the remoteness of the area and many of its national parks. Yet, significantly, each different grouping of people is vying for control in some sense—over people, animals, or the land itself.

Despite large swathes of land being transferred into National Parks in recent years, conservation outcomes for the region remain poor (Commonwealth of Australia 2020; Department of Environment and Science 2021b). Cape York echoes the international situation, which is that international goal setting and agreement making, as well as an increase in protected land around the world, has failed to translate into markedly improved outcomes in terms of biodiversity and conservation (IUCN 2022; Secretariat of the Convention on Biological Diversity 2020). Conservation regimes designate some forms of care for land as appropriate, while excluding and ignoring others. Significant shifts in recent decades have seen developments in the space of comanagement, and acknowledgment of the environmental knowledges and practices of Indigenous peoples around the world as valuable (Recio and Hestad 2022; Convention on Biological Diversity 1992). As Gomeroi scholar Heidi Norman and colleagues point out (2022), the Indigenous land estate in Australia is significant in terms of both geographical area and biodiversity values. However, Indigenous environmental governance, through structures like the formal joint management program that exists in various Australian national parks, presents new challenges and complexities alongside opportunities for Indigenous communities (Carroll 2015; Nadasdy 2003; Reardon-Smith 2024, 2025).

As this book will demonstrate, the constructed binary between nature and culture comes unstuck in the day-to-day and quotidian ways in which people relate to environments and protected areas. In Cape York, people make their lives and livelihoods in and through their relationships to land, and the categories through which environments are apprehended in conservation are frequently confounded through everyday practices of making do. People find new ways to live in and with environments, in the contexts of economic marginality, land tenure changes, and ecological pressures, including the ongoing spread of invasive species and the effects of a changing climate. Aboriginal Traditional Owners, government-employed park rangers, and cattle graziers alike engage in caring labor to look after land, and their own livelihoods, making necessary compromises to engender workable landscapes and livable lives.

People, Place, and Methodology

Despite her initially abrupt attitude, Pam, the grazier I described earlier, was kind to me, agreeing to let me visit. When I arrived in Cape York, in May 2018, Pam was the first person I stayed with. In 2018, the main road through Cape York—the Peninsula Development Road (colloquially called the PDR) was paved only until the town of Laura. North of Laura, the road was dirt. In the dry season the dirt road develops deep corrugations and dust holes, and in the wet season the softer sections become slippery, muddy expanses, impossible to drive through without a four-wheel-drive vehicle. The turnoff to Pam's property is a couple of hours' drive north of Laura. In between the small town of Laura—which consists of a cultural center, a pub, a campground, a roadhouse, a small school, and, of course, a rodeo ground—and Pam's property, there isn't much. A few river crossings, a couple of roads that intersect with the PDR, and a roadhouse with a lone, domesticated emu wandering around the cars and customers.

It is largely because of Pam's generosity that I was able to work with cattle graziers in the region. Some graziers were initially reticent to be involved in my research, having had negative experiences with researchers in the past. One grazier, Alan, scrutinized my credentials carefully as he slowly ate his lunch (a corned beef sandwich) and I sat awkwardly beside him on his front porch. Alan was a brusque man, dressed in faded and torn workwear, always barefoot, and missing some of his front teeth from horse and bull-riding mishaps. He told me that he wanted to make sure that I wasn't an "undercover greenie." It took a long time for Alan to warm to me. However, with Pam's blessing, he and several other grazing families agreed to be involved in the project. Eventually, Alan came to trust me—grudgingly and contingently.

Several facets of my personal background made building rapport with these grazing families possible: being young, a woman, settler-descended, and, most importantly, having grown up on a rural property in regional Queensland. Many of these graziers were puzzled when I explained that I didn't want to merely sit and talk with them but also wanted to take part in their daily work activities. I gradually earned the respect of the graziers by engaging in mustering, yard work, fencing, cooking, cleaning, and by not complaining about the insects, dirt, heat, humidity, early mornings, and long days.

While, from the 1970s onward, a number of pastoral leases in Cape York have been purchased by multinational companies, the graziers that I worked with all self-identified as "pioneers." All were either descended from or married into one of a handful of established grazing families in Cape York, families that had lived in the region for several generations. I worked closely with four different stations and more loosely with others, and each of these families was entwined with the others through blood relations, historical business partnerships, or multigenerational friendships. While most of these graziers identified as White, at least one grazier acknowledged her Aboriginal ancestry.

The properties that these graziers live on are relatively small scale compared to cattle stations elsewhere in northern Australia, all supporting fewer than 2,000 head of cattle. Their business operations are small, and the work is generally carried out by a husband-and-wife team with occasional assistance from family members, neighbors, and contract musterers. The grazing industry in Cape York has dwindled in recent decades due to climatic and market fluctuations, and the fact that Cape

York is decidedly marginal grazing country and geographically distant from markets. As noted, without unpaid Aboriginal labor, many stations became unviable. Today, most graziers have diversified their businesses and take part in a range of economic activities, including road building, truck driving, gold mining, and tourism. Most small-scale graziers in Cape York are compelled by market pressures to sell their cattle for live export to Indonesia, a practice that remains controversial in Australia. In addition, the region's proximity to the Great Barrier Reef contributes to a public perception that grazing cattle in Cape York is environmentally unsound. These factors go some way toward contextualizing why many graziers were hesitant to allow a researcher into their lives, workplaces, and homes.

My route to building relationships with the Aboriginal Traditional Owner groups and rangers was more prescribed. I initially contacted an Aboriginal ranger group called Rinyirru Aboriginal Corporation once I arrived in north Queensland, as I was aware they comanaged the largest park in the region. I met with the Rinyirru Aboriginal Corporation senior ranger, Peter, in the corporation offices in Cairns, a small building set in a ground-level block of shops, just off the highway that runs through the middle of the tropical city. Alternatively energetic and enthusiastic, and exhausted by the enormity of his role and the teething issues with the joint management project, Peter was enduringly friendly and open toward me. He took his role as a cultural person seriously and spoke at length about his discussions with his "old people" and the significance of the knowledge he gained from relatives, in particular, his uncle. He often told me that his responsibilities were along two lines that sometimes coalesced and sometimes came into conflict. There was his role as a manager for the corporation, which involved negotiating with his Queensland Parks equivalent and mediating between Queensland Parks and the board of directors for the corporation, and then there was his role as a cultural person, and a custodian for his ancestral homelands— some of which fell within the bounds of the park. During my time in Cape York, Peter was under enormous pressure, related to budget constraints and managing the different expectations of the families he represented and the government organization he worked in partnership with. Yet, he often made time for me and assisted me throughout my field research, providing advice, maps, and phone numbers of possible contacts.

Rinyirru Aboriginal Corporation is composed of the native title-recognized Traditional Owners of Rinyirru National Park (formerly Lakefield National Park), the Lama Lama and Kuku Thaypan peoples, and the Bagaarrmugu, Mbarimakarranma, Muunydyiwarra, Magarr-magarrwarra, Balnggarrwarra, and Gunduurwarra clans. Rinyirru National Park is the second largest national park in Queensland, with an area of 5,370 square kilometers (roughly 2,073 square miles), and is a popular destination for campers, bird-watchers, fishers, and folks traveling in their trailers. It is home to wetlands, woodlands, savannas, grasslands, coastal estuaries, and mangroves. For most of the time I spent in Cape York, Rinyirru Aboriginal Corporation had a reasonably small workforce, with only five rangers.

While Peter was mostly based in the office in Cairns and dealing with a variety of administrative and management tasks, the other Rinyirru Aboriginal Corporation rangers lived on and off in the national park. These rangers were all from families that had both ancestral and more recent historical connections to the park through family members who had been employed as stock workers at Lakefield Station before it was purchased by the state government and declared a national park in 1979. These families all had stories of relatives being forcibly removed from the site of Lakefield Station to missions across north Queensland, including Hopevale, Yarrabah, and Palm Island. While Native Title was recognized over Lakefield National Park in 1997, it took a further fourteen years of negotiations between Traditional Owners and the Queensland state government before the park was handed over to Aboriginal control. As with many native title determinations, this was a long and sustained struggle, and many of the senior elders who labored for years on the land claim died before the land was handed back. In 2011, Lakefield National Park was renamed Rinyirru National Park and formal joint management of the park between Rinyirru Aboriginal Corporation and the Queensland Parks and Wildlife Service was instituted.

Through my connections with the Rinyirru rangers, I was introduced to the chair of the Lama Lama Land Trust, a cheerful, generous, and pragmatic man named Kevin. Kevin also sat on the board of the Rinyirru Aboriginal Corporation, and hearing about my research, invited me up to Lama Lama Country to meet the rangers he works with and trains. This group, which jointly manages Lama Lama National Park, a much smaller park to the north of Rinyirru National Park, employs around fifteen rangers full-time year-round and has a fluctuating workforce of around forty casual rangers. Somewhat unusually for the industry, the Lama Lama Land Trust has a fairly equal number of men and women employed and has a large number of women in leadership roles. Lama Lama National Park was the first joint-managed national park in Queensland, established in 2008. While a public road runs through the park, there are no facilities and little visitor access. Most of the Lama Lama Land Trust rangers live between the central Cape York town of Coen and Port Stewart–Yintjingga, a small township north of Princess Charlotte Bay.

Historically, there have been several geographically distinct Lama Lama communities. Most of the rangers who were employed at the time of my fieldwork were descended from the Port Stewart–Yintjingga Lama Lama people. The Port Stewart–Yintjingga community was inhabited continuously until 1961, when the entire community was forcibly removed to a mission farther north called Bamaga and, later, Injinoo in the very north of the peninsula. There were attempts made by the grandparents of many of the current Lama Lama rangers to return south to their home, but these all ended in police capture and transport back to Injinoo. It was not until the early 1990s that the Lama Lama people were able to reestablish a small outstation community at Port Stewart–Yintjingga. It is here that the current ranger base is located and that some of the rangers live. Through a long and sustained struggle, the Lama Lama community gradually gained title over multiple parcels of their land in tenure types including National Park, Aboriginal freehold, and pastoral lease. I also worked with Queensland Parks workers employed at various levels, from management to specialized pest teams and operational rangers who live full-time in the national parks. While most of the operational rangers heralded from Queensland, few rangers had previous personal ties to Cape York. The exceptions are those rangers who have filled Indigenous-identified positions, many of whom are Traditional Owners for the region. Some of the rangers had enjoyed long careers with Queensland Parks, whereas others had moved into these roles toward the end of a career in another industry, like mining or pastoral station work. Generally, operational rangers came to Queensland Parks with practical, hands-on skills like fencing and mechanical expertise, which are considered highly valuable in remote national parks by Queensland Parks management.

The ranger in charge for Rinyirru National Park at the time of my fieldwork was a man I'll call Ray. Ray had the kind of sun-pickled skin shared by most White men in north Queensland who have spent their working lives outdoors. His background was in heavy machinery and road building, for which he was employed by Queensland Parks in the Cape for over two decades before moving into his current role. He was the kind of ranger who saw "real work" as what could be done outside, although as the ranger in charge a large proportion of his work was administrative and office based. Known among his staff for his wide range of catchphrases (most notably, "AG?" meaning "all good"? and "PNG" meaning "proper no good"), Ray was at once instantly likable and deeply divisive. Aside from one Indigenous ranger and the wife of an operational ranger who did administrative work a couple of days a week, all of the Queensland Parks rangers employed in the park during my fieldwork were men, reflecting a tendency in remote-area ranger work. However, among the off-park management staff, there were many women, mostly with college degrees in conservation biology or ecology.

Throughout the duration of my fieldwork (around sixteen months in total, with substantive fieldwork occurring in 2018 and 2019 and shorter trips in 2017 and 2020), I spent work and leisure time with my research participants. I took part in a variety of land management activities: mustering, fixing fences, spraying weeds, planting trees, mapping fire scars, monitoring climate change impacts, and undertaking controlled burns, among other things. I also helped cook and clean, prepared meat for domestic consumption, went fishing and bird-watching, fed poddy calves and goats and chickens, and drank seemingly endless cups of tea and lukewarm cans of mid-strength beer. I went to Cape York with certain questions in mind, but was led in other directions by what the people I worked with found to be important.

It's important to note that throughout the book I refer to Aboriginal Traditional Owners, Aboriginal ranger groups, graziers, and Queensland Parks rangers collectively as "land managers." I have reservations about the term "land managers," largely because, as Howitt and Suchet-Pearson (2006) argue, the concept of management relies on a separation of nature and culture that is distinct to a Western ontology

(Descola 2013) and implies an assumption that this separate nature can be mastered, controlled, and managed by humans. Thus, the term "land manager" fails to encompass how many Aboriginal people conceive of and practically engage in caring for their ancestral homelands. However, in recent work (Gammage 2012; Pascoe 2014) "land management" has been applied to Aboriginal forms of caring for Country, and the use of this term represents an important political stance that is related to Aboriginal sovereignty. My sense is that "land manager" is an imperfect term and I use it with caution.

Shape of the Book

In this book, I explore how environmental knowledges and ethics are coproduced in intercultural assemblages and in counterintuitive and unexpected ways. The text is organized into three main themes: wrangling (chapters one and two), seeping (chapters three and four), and engulfing (chapters five and six).

The first section of the book details processes of domestication in Cape York and the sprawling impacts of these processes. Used within the pastoral industry to describe attempts to manage, control, or bring into line unpredictable animals, like bull catching, "wrangling" points to the work involved in trying to assert control, as well as the presence of the risk that such attempts may go awry. A kind of domestication, I use this term to evoke the domestic livestock that populate the pages of chapters one and two; cattle and the infrastructure used to manage and control them. However, more broadly, I use wrangling to gesture at the project to domesticate this still-remote region of Australia through the pastoral industry (animal husbandry) and—more recently—attempts by the Queensland Parks and Wildlife Service to manage and tame the unruly pastoralists whose leases neighbor national parks.

In chapter one, I investigate the relationship between people and cattle, exploring how this species has come to mediate how some Aboriginal people and settler-descended graziers understand the land and their relationship to it. The intimate relationships between graziers and their cattle encompass both instrumentality and care, and such care is echoed by (particularly older generations of) Aboriginal people who continue to see cattle as belonging in the region in some sense and continue to see cattle work as a "proper Aboriginal pursuit" (Smith 2003a). It is through cattle that many people continue to make do in Cape York. However, cattle are no longer the primary way through which people come to relate to land in the region, or make a living. I follow these changes and emerging tensions as the pastoral industry gives way to so-called "green collar" jobs in conservation and land management.

Land tenure changes have transformed the social make up of Cape York, but the introduction of boundary fences, too, has had profound material and symbolic impacts. Chapter two traces the history and contemporary social role of boundary fences, bringing into sharp relief some of the disagreements and disputes between graziers and Queensland Parks. Cattle (both feral and graziers' stock) are a significant problem for Queensland Parks. They enter parks through broken boundary fences and floodgates, and within the boundary of the park, cattle undergo a discursive shift from "livestock" to "pest species." The movement and subsequent management (and management failures) of cattle provide insight into ideas around ownership, boundaries, and fences. Nowadays, boundary fences and Queensland Parks' cattle management strategy lead parks to see graziers and cattle alike as unruly, as tussling, as refusing to be organized and managed. They are (both) characterized by Queensland Parks as willfully antagonistic, illuminating the extent and limitations of Parks' management reach.

The second section of the book, "Seeping," attends to how land managers and locals live with and against invasive plant and animal species. Chapters three and four describe the ways in which species are classified and interpolated, with ramifications for how they are managed and understood by the people of Cape York. While the term "entanglement" has found favor in recent years, I have chosen to use "seeping" to think about the uncomfortable nature of the porous boundaries between things. Seeping is often unstoppable and not always welcome; it evokes the kind of movement that frequently demands attempts at containment and control. In this section of the book, I seek to take seriously the material impacts of invasive plant and animal species, and the efforts toward their management, rather than conceptualizing invasives as presenting new possibilities for flourishing.

Chapter three introduces the notion of killing as a form of care to discuss weed control. Weed control comprises much of the land management that happens in Cape York—particularly during the wet season. The way that differently located people classify invasive plant species as more or less problematic reveals much about what kinds of landscapes— and forms of landscape care—are considered preferable for different land managers. All land managers engage in weed control, albeit for a variety of reasons, and thinking with weeds brings together issues of inheritance and responsibility, as well as purity and hybridity. Seeking to control weeds is also illuminating for thinking about scale and temporalities; to enact killing weeds to care for landscapes, practitioners must attend to how plants behave. Their caring labor is constrained by the mismatch between phytotemporalities, to borrow Margulies's (2023) term, and the rigid demands of funding cycles and project timelines.

A key pest species targeted for control by Aboriginal rangers, graziers, and Queensland Parks alike is the feral pig (*Sus scrofa*). Pigs damage sensitive wetlands, threaten endemic species, damage important story places, and churn up mud around water sources, presenting a risk to cattle. Chapter four investigates how feral pigs are categorized and controlled, with a particular focus on how killing pigs can function as a form of (violent, compromising) care for the land. However, while all land managers kill pigs to a greater or lesser extent and for a variety of reasons, Queensland Parks also devotes time and resources to stopping the illegal hunting of pigs within park boundaries. I investigate *who* is allowed to be killing as care, and whose killing is seen as uncaring and problematic. Forms of authority structure how the killing of pigs is framed—sometimes as conservation, sometimes as antisocial and illegal behavior.

The final section of the book, "Engulfing," moves away from a consideration of nonhumans at the species level to thinking about elemental nonhumans: fire and water. These are elemental forces that emerge in peoples' lives in ways distinct from the other nonhumans that land managers are grappling with; they represent large, sweeping, landscape-level impacts. Fire and water both, in various ways, generate implicit and explicit considerations of the changing climate. By using the term "engulfing," I highlight how things like fire and water are, in some ways, considered to be nonhumans that operate outside human control, even as their impacts are experienced at the local level, revealing values, beliefs, and tensions around how people believe they ought to prepare and recover from related wide-ranging impacts. I dig into the tension between people adopting a seemingly uncaring position toward climate change and the actual caring labor they do to prepare for and mitigate its effects.

Fire, both wild and controlled, affects all land managers in Cape York. Chapter five details the different types of fire—cool burns, storm burning, carbon sequestration, and wildfires—and explores how each type of fire encourages forms of critique between and among the various land managing parties in Cape York. Fire knowledge, originating with Aboriginal Traditional Owners, is now interculturally mediated and adapted to serve diverse purposes and achieve particular outcomes. Through examining the burning practices and perspectives of Aboriginal Traditional Owners, Park rangers, and cattle graziers, the ideological underpinnings of different fire regimes emerge. These insights disrupt some of the accepted wisdom around fire management and cultural burning in Australia. In both intentional burning and bushfire, many people frame the actions of their neighbors as uncaring—in applying too much or too little fire, or in fighting fires in a way that can be perceived as protecting one's own interests at the expense of their neighbors.

Water also shapes the social world of Cape York. The temporality and variability of the monsoon provides insight into how weather events come to be understood in various ways by different people. In chapter six, the monsoon emerges as a way to think about local responses to climate variability and climate change. Where some Cape York residents reject or are ambivalent toward the explanatory framework of climate change, preferring to frame variability in terms of "natural cycles," other land managers actively take part in climate change monitoring projects and use the language of climate change with ease. Taking attitudes toward the concept of climate change as my starting point, I think through how a seemingly uncaring position is undercut by and through caring labor, exploring the gap between a stated politics and deeply held and contradictory environmental values.

In the concluding chapter, I explore the relationship between a single cattle grazier—Pam—and the endangered golden-shouldered parrots that she works to protect. Pam has developed the kind of "response-ability" and art of noticing that Haraway (2008) and Tsing (2015) speak about, and her deep and enduring care for the parrot is evident in her words, actions, and attunement. Yet Pam is caught in a contradiction—

her life and livelihood is reliant on cattle, but she knows and acknowledges that cattle are the worst thing for the parrot. By thinking with and
alongside Pam and the parrot, I explore what happens in a space where
care mingles with complicity. Pam does the messy and entangled work of
caring for the parrot, in awareness of her own culpability in harming the
parrot. Her conservation work is impure and imperfect but is enduringly
grounded in a kind of deep and steady love; a space that allows for messy
knots to be made and unmade.

Despite the contradictions and tensions that will emerge throughout
this book, a deep ethic of care for the environment and the region (peopled by humans, plant and animal species, and ancestral echoes) is the
thread that links (contingently, contextually, tentatively) all those who
enduringly live and labor on the land in Cape York, all of those who
make do.

PART I

Wrangling

Cattle being mustered, Hillview Station, 2018.

ONE

Cattle

Early in the morning, on a warm and dry day in August, I was in an old utility vehicle with cattle grazier Pam and her two grandchildren. We were slowly making our way along a bumpy bush track to a mustering camp in the northern section of Pam's family's pastoral lease, moving through sandy scrub and stands of melaleuca and acacia trees. The men in the family, along with a British backpacker working at the station and three contract musterers, had departed earlier that morning to prepare for moving a large mob of cattle from this region to the yards located near the house. By this time of the year, Pam and her family had already completed much of the mustering. On this day, the terrain to be covered was relatively accessible, and Pam's husband had, somewhat grudgingly, agreed to let me come along. Learning that I was to participate, Pam's grandchildren—ages seven and nine—had successfully argued that they should be allowed to miss their homeschooling lessons on this day and participate too. Because she had grown up taking part in cattle work alongside her brothers, their mother agreed.

When we arrived at the mustering camp, the graziers and stock workers there had already gathered the cattle into a mob, ready to be moved. In the past, graziers in Cape York would muster on horseback, taking several weeks to muster each section of their leases as the cattle were dispersed across vast distances. Nowadays, graziers use a combination of helicopter mustering, "trap paddocks" or "spear traps,"[1] which draw cattle into centralized locations, and all-terrain vehicles, referred to by locals interchangeably as quad bikes and four-wheelers. This makes the mustering faster, more efficient. Some graziers complain about the negative effects of these changes, complaining that cattle nowadays are less domesticated than their forebears due to less time spent with humans, and asserting that the helicopters and bikes push cattle to walk too fast, losing condition. In a lot of ways, these critiques of changes to the rural industry can be read as graziers mourning the loss of a valued way of life and an identity wrapped up in horsemanship. But the economically marginal status of grazing in the region (Neale 2017) means that efficiency, in terms of time, labor, and resources, is key.

Having had some brief lessons from various graziers in how to safely ride a quad bike, I set off on the bike belonging to Pam's grandchildren—the only bike available on this day. One of the grandchildren rode with Pam, the other with Pam's husband, Mike. I was part of the tail end of the mob so, along with Pam and the British backpacker, I slowly rode my bike behind the large group of cattle, gently nudging them along. We were pushing the cattle from a trap paddock in the north of the lease to the yards at Hillvale, where they would be processed, vaccinated, treated for parasites, and some cattle would be prepared for sale, intended—as most of the cattle in Cape York are—for live export to Indonesia. As we rode, it became clear that everyone was holding a specific position in relation to the cattle. Pam, Mike, the British backpacker, and I were at the back; two of the contract musterers kept the sides of the herd in check, while a third, Gina, circled the mob, guiding stray cattle back into the throng; and Pam and Mike's son, Justin, was in the lead. Gina had her dogs with her—three eager and skinny working dog crosses, a mixture of kelpie, cattle dog, and border collie. They mustered alongside her, responding to her calls and commands, leaping on and off her bike and nipping at the heels of stray cattle.

The cattle marched slowly, with beasts only occasionally splintering off and requiring a flurry of activity from the stock workers to bring them back under control. The preferred breed in Cape York is the brahman, a breed that originated in India and is known for handling the hot and humid conditions of the tropics with more ease than European breeds. I was told by one grazier that brahman are the only breed of cow that can sweat, that their skin is black so that they don't burn under the harsh sun, and that they are better "walkers" than other breeds. The cattle did—for the most part—undertake the journey in an orderly fashion. Dust swirled around the herd, and there was the constant sound of the cattle calling for each other. On occasion, a small calf would become separated from its mother, who would cry out, panicked, until they could be reunited. Some mothers and calves were permitted to peel away from the herd, the graziers and stock workers acknowledging that the walk was too far for such young cattle. It was late in the day when we arrived at the yards. The cattle were let into a large paddock called "the cooler" to rest, feed, and water. They would be processed the following day.

Relating to Cattle

Cattle exist in Cape York as both domestic livestock and feral animals. They serve a variety of purposes: as a key component of most people's diets, as livelihood, and as a link to a valued way of life. Cattle simultaneously occupy the discursive positions of a symbol of life and livelihoods, of economic opportunity, and of a problematic pest animal responsible for environmental degradation. They are found on pastoral leases and in national parks alike. Despite the fences and deep rivers crisscrossing the region, cattle manage to wander into all sorts of geographies. In Cape York, cattle are "good to think with" (Lévi-Strauss 1962), as Cape York land managers' relationships to land—and each other—are frequently mediated through cattle. This chapter is about cattle, but more broadly is about how, through cattle, people have attempted to wrangle, persist, and make do in this economically marginal place. I trace the ways that people make a living, and make a life, in Cape York. These efforts to make do overlay and mesh with the still-present colonial history of this place, the wrangling of people and place that led to Cape York being remade as a site for cattle grazing.

Graziers have deep, intimate, and multiple relationships with cattle. Cattle shape how graziers manage their land and even how they move across the landscape. The ways that graziers interact with cattle are sometimes brutal and sometimes tender, oscillating between a gentle kind of empathy and attention-paying and moments of frustration, physical violence, and enforcing control. Likewise, Aboriginal people have had and continue to have close relationships to cattle as an economic resource and food source, as well as an important cultural marker. Particularly older Aboriginal people in Cape York who worked as stockmen, or ringers,[2] in their youth display what Baker (1999) has called a "cattle identity." While Queensland Parks as an institution categorizes cattle in Cape York as a pest animal that needs to be excluded from national parks and areas with high conservation values, in local framings cattle emerge as something that has come to "belong" in the landscape. That is, to graziers and Aboriginal Traditional Owners, the presence of cattle does not necessarily indicate that Country is being poorly cared for.

In the yards, cattle are generally gathered into the largest pen. Here, the cattle are given time to feed and water and settle overnight. The next day, bit by bit, the cattle are processed. Smaller groups of cattle are brought into increasingly smaller pens, until the cattle are lined up in single file in a kind of chute called "the crush." In the crush, the graziers work quickly to process the cattle. They are injected for botulism, a bacterial disease, and each beast that is vaccinated has the long hair at the end of their tail trimmed to indicate that they have had their vaccination for the year. This practice is called "bang-tailing." While in the crush the numbers of male and female cattle are recorded. If any animal in the crush has some kind of physical issue—for instance, the cows sometimes experience prolapse after giving birth in the scrub—this is dealt with swiftly. Prolapses are reversed and sewn up by dexterous graziers in a matter of seconds. On most occasions, the cattle are then pushed through a deep channel of liquid dosed with insecticide called "the dip." This is a deep concrete trough into which the cattle must leap or fall, swim through, and then clamber out a concrete slope on the other side. The purpose of the dip is to protect cattle against ticks. Sometimes the cattle are merely sprayed with "drench"—antitick fluid—while standing in the crush, but graziers prefer the dip because it tends to be more economical. The cattle respond to the dip in various ways. Older cattle, obviously recalling having done this before, jump, somewhat gracefully, into the dip. One grazier commented to me that the "old girls" know if they make a good jump, they only need to be in the dip for half as long.

For the calves though, the dip is an unknown entity. Sometimes four or five calves are bunched up, trying to scramble away from the precipice, until they tumble, one after the other, into the liquid.

Calves generally face a more traumatic experience than their parents. Separated from their mothers for the first time, the calves are moved through a smaller crush and caught in a large metal contraption called a cradle, in which they are constrained. While in the cradle the calves are branded, ear-tagged, de-horned, and the males are castrated. This whole process is completed very quickly, over the time frame of about two minutes. At Pam and Mike's station, I watched four experienced stock workers methodically carrying out these procedures on the calves. If the calf was a male, one person would hold their legs still while another, scalpel in hand, quickly removed their testicles. As this was happening, another stock worker would cut the station's mark into the calf's ear and then dig out the roots of any burgeoning horns with a blade. Finally, the large metal brands, heated with a portable gas flame, are imprinted into the calf's rump, stamping it with the station's brand and a number indicating the year that it would be old enough to go to market. Throughout the ordeal, the calves bleat pitifully, presumably in pain and panic. When released from the cradle, the calves form a small huddle together in the edges of the pen, seemingly disoriented. The smell in the yards is intense: an acrid mixture of dust, chemicals, blood, manure, and burned hair.

On another occasion, I watched a grazier spaying grown female cows. He told me that he did this to stop certain cows becoming pregnant during the harsher times of the year, when he feared a pregnancy would likely result in death for the cow and calf. The spaying was difficult to watch—occurring without anesthetic, the grazier would slice an incision into the side of the cow before reaching in to remove her ovaries with a small blade. He would discard the ovaries, allowing his small Jack Russell dog to eat her fill. He would then shake some antiseptic powder in the wound, sew up the gash with brown string, spray it with another antiseptic liquid, and then send the cow on her way. This manner of spaying cattle is illegal as the relevant authorities require a professional veterinarian to carry out the procedure, but this is not viable on a small cattle station eight hours' drive away from the nearest veterinarian.

Most small grazing operations in Cape York only prepare steers (cas-

Poddy calves, Apollo Station, 2018.

Mustering, Hillview Station, 2018.

trated males) for sale. The reasons for this are largely economic; female cattle must be pregnancy tested before going to the sale yards, and this is an expensive enterprise. Alternatively, some graziers spay female cattle so that they can go to market, but most stations I worked with preferred not to do this, with spaying female cattle for sale considered a last resort, turned to only when trying to reduce stocking rates dramatically.

For a beast to go to market, there is a requirement that it is either "de-horned," as described, or has its horns "tipped." Tipping horns involves cutting off the sharp, pointed ends of the horns, and this is done to reduce the risk of animals locking horns or injuring each other when in close quarters—either at the sale yards, meat works, or the boat to Indonesia. Under the legislation, any animal over three months old should have anesthetic administered before de-horning or tipping, but for many graziers working on remote stations this is considered to be an unnecessary expense. As one grazier put it to me, "we don't go out of our way to hurt animals, but it's just some of the things you have to do to present them to market. I think you're supposed to have them de-horned before they're three months old . . . but a [in] a lot of areas it's unpractical because you don't find them."

These kinds of physical interactions with cattle indicate an interspecies relationship that is characterized by instrumental value and control. Such a human-animal relationship emerges from the grounds of capitalism and anthropocentrism, creating what Wadiwel has called a "bad infinity" (2023, 9–10). However, as Haraway suggests, inherently unequal and instrumental relations between humans and animals do not necessarily preclude the possibility of an ethical relationship (2008, 71). If, within instrumental relations in which animals' bodies and labor are used by humans for research or food production, humans are able to come "face to face" with animals and recognize not only their suffering, but their status as subjects rather than simply objects, then some element of care and what Haraway calls "response-ability" can exist (2008, 74–76). Speaking of dogs in historic laboratory experiments, Haraway writes, "to share the dogs' suffering . . . would be not to mimic what the canines go through in a kind of heroic masochistic fantasy, but to do the work of paying attention and making sure that the suffering is minimal, necessary and consequential" (2008, 82).

The graziers and stock workers carry out this work of paying attention, which has also been framed variously as an "arts of attentiveness" (van Dooren, Kirksey, and Münster 2016) and the "art of noticing" (Tsing, 2010, 192–94). The procedures they inflict on the cattle can be considered in some ways to be necessary and consequential. They are sometimes, though not always, performed for the animal's well-being and frequently in order for the cow to fulfill the grazier's requirement that it be salable. Aside from the lack of anesthetic used, the suffering inflicted during the procedures is also minimized as much as possible through efficient systems. Graziers are cognizant of bovine suffering. The grazier I watched perform the spaying operation admitted to me that he didn't enjoy doing it at all but considered it part of his duty in ensuring the long-term well-being of a cow who may not survive another pregnancy. This can be read as a noninnocent act of care, composed of empathy, violence, control, and worry (Martin, Myers, and Viseu 2015). This grazier spays cows for multiple reasons, sometimes to ensure that they survive the end of the dry season, when there is little feed available, sometimes so that the cow can be sold, and sometimes because a particular animal has undesirable traits that the grazier did not want passed on to descendants. Spaying the cow, then, can be thought of as an act of contingent and partial care, demonstrating care and concern for both the well-being of individual cattle and the well-being of the herd at large, but—as always—in a way that is deeply entwined with the economic relationship between humans and their cattle. This is a form of care that is not purely benevolent, and not purely self-interested, but characterized by ambivalences.

While much writing on the coexistence of violence and care in human-animal relationships focuses on acts of violence that function to protect or prolong the life of a species (violence for biodiversity, violence for conservation) (van Dooren 2014, 2019, 2011; Bocci 2017), this mingling of violence and care is different. Instead, I find more alignment with how the concept of attunement has been used by Haraway (2008) in discussing lab animals, and in Blanchette's (2020) work on industrial hog farms in the American Midwest. In these instances, the animals play a sacrificial role, but this does not preclude the workers who labor with and on these animals from learning to pay attention to the lived reality of these animals, seeking to minimize their pain and discomfort through a kind of intimacy. As Blanchette (2020, 116) demonstrates, this intimacy—or, in the case of one worker, a refusal of intimacy—is constitutive of the workers themselves. They are, in some sense, remade into a particular kind of productive worker on the factory floor, attuned to specific (and relevant) hog characteristics. Similarly, in perpetrating forms of violence toward cattle, culminating in their inevitable slaughter in Indonesia, alongside material acts of care, graziers are constituted through this particular form of human-animal intimacy. The relationship is simultaneously caring, mundane, and exploitative. This human-cow intimacy is at the heart of the graziers' valued way of life; it is their reason for being in the region. Their relationships to people, place, and livelihood are all mediated through cattle. And yet, relationships between humans and cattle in Cape York are very much shaped by economic forces. At times intimate, human-bovine interactions remain textured by the instrumentality of the relationship.

Selling Cattle

In recent years there has been an increase in requirements around biosecurity and traceability in Australia's cattle industry. In the past, cattle were identified through a tail-tag with each station's unique property number. However, technological developments and concerns around biosecurity have led to this system being abandoned in favor of an electronic ear tag called a National Livestock Identification System tag (Meat & Livestock Australia 2007; Queensland Government 2022). The benefit of this kind of electronic tag is that it can hold data about the various places that cattle have been, so it is possible to trace their route from birth to sale. This is relevant in situations where cattle are contaminated with disease, enabling authorities to determine where contamination may have occurred. Such biosecurity measures emerged in the wake of a bovine spongiform encephalopathy disease outbreak, in which bulls from a particular stud in the town of Rockhampton were sold to various stations in Queensland and the Northern Territory, spreading the disease (McCarthy et al. 2014). In this instance, the diseased stock had to be destroyed. Graziers have adapted—somewhat grudgingly—to these changes, accepting that they need to get their heads around technological advances and be prepared for government agents to audit their properties to check for contaminants and correct protocols. Many graziers lament that running cattle is, nowadays, much more "like a business" than it used to be.

As grazier Alan said to me, "at the end of the day, that's what Australian meat is known for. It's clean and green. Because of the traceability of it. I mean, it's a pain in the arse, but I guess that's the benefit from it, put it that way."

Many Cape York graziers consider the region to be "good, safe, breeding country." It's country that's good for breeding cattle, but not for growing them. Those who have the access and money available send their cattle to fattening blocks farther south, in the Daintree region, or on the Atherton tablelands. Here, the cattle put on weight and, as such, can fetch a higher price when they go to market. These fatter cattle may go to the meat works and end up being consumed domestically. But for many Cape York graziers this is not an option. Their cattle, straggly and skinny, are destined for live export to Indonesia (via the sale yards in the town of Mareeba), unable to be sold profitably within Australia without the value-adding that access to a fattening block can provide.

However, even with access to a fattening block some graziers still find live export to be the only viable option. When I asked where their cattle go after a couple of months putting on weight in the Daintree, grazier Diane replied, "Export. Ours go to export because they're so small. We don't have enough area, really." Her husband, Bill, explained that cattle going to the meat works have to be around 450 kg (992 lb.) in weight, whereas cattle for live export can be as small as 260 kg (573 lb.) in weight.

"Well, we never really have enough for meat works, so we usually sell them through the sale yards. That's how you keep your herd moving," Diane told me.

As well as weight restrictions, the "quality" of beasts determines whether they are appropriate for domestic markets or live export. While some graziers—like Alan and Bev—work hard to curate a good, high-quality herd, this is less of a concern for others. Alan explained to me that the industry enables this, because live export boats will take "pretty much anything," even cattle that Alan would deem to be "rubbish" stock.

When there were more cattle properties in Cape York, graziers sent their cattle to be loaded onto boats in Weipa on the west coast of the Cape. Alan recalled to me that when he first took over his family business after the sudden death of his father in a mustering accident, some thirty-five years prior, there was enough cattle going to Weipa to necessitate three boats a year. At this time, the various graziers in the region banded together to organize the boats. However, it is no longer profitable or even viable to do this. As Alan told me, "Now you'd be lucky to load one [boat]. It's a business sort of decision we make. Like, we wouldn't support a boat if the boats went this year, because we haven't got the stock here to do it." Alan, along with all the graziers I worked with, sent his cattle to the sale yards in Mareeba that year.

However, sending cattle from Cape York to Mareeba also carries significant expenses. Pam explained to me that for the truck to cart the cattle to Mareeba, it ends up costing about AUD$50 (approximately USD$32) per beast. "Probably more sometimes," she reflected. "And then you've got all them other fees on top of it. You've got yard fees down there, and hay if they're fed hay. Then you've got dipping. . . . There's about four or five fees, I was looking at them yesterday when I was doing the tax. You know, it costs a lot to sell a beast in Mareeba, so if you live a long way away and the road's really rough, you've got to pay big money to get them there."

Given that they are all quite small operations, generally running around 2,000 head of cattle per station, stations in Cape York deal with stock agents, rather than having a direct relationship with exporters. At Pam and Mike's station, I observed the stock agent sorting the 120 or so steers that they had ready for sale. Along with Pam and Mike's grandchildren, I sat on the top rail of the yard fencing, watching as the agent designated each steer as a number 1 or number 2 beast. The number 1 steers were those that the agent believed would fetch a higher price; they were larger beasts and included all the white steers in the herd. One of the contract musterers mused that she was unsure why the white cattle were worth more than the reds, telling me that she personally believed the red cattle were "better eating." Once the cattle were sorted, they were let gradually into a narrow chute and then encouraged onto the waiting truck with a mixture of whistles, sounds of "ch-ch," "pshhh," and "aarggh!," slaps of the rump, and quite a few hits of the electronic cattle prod (colloquially called the "jigger").

Later, Mike told me a little about their stock agent, Jim. He told me that Jim is contracted through Landcare and Cape York Natural

Resource Management (CYNRM), and that Mike has been very happy in his dealings with Jim. He explained that he and Pam used to use Jim's nephew, Rocky, as a stock agent, but that Rocky was unable to sell Mike's poorer stock. Jim, on the other hand, had a reputation for being able to sell anything. Mike said that he had not known why Jim went out of his way to help Mike and Pam shift their poorer quality cattle, but that someone had told him that Jim does it mostly to "piss off" Rocky. "People say he drinks with the buyers," Mike said to me. "But I don't care."

To continue to grow cattle in Cape York, graziers are generally compelled to take part in live export. Before they even get onto the boat, cattle are transported long distances along rough roads. Their journeys on live export ships can take between seven and twenty-three days. The economies of scale necessary to make live export financially viable mean that the journey is not a comfortable one, and that it is accepted that a significant amount of animals will not survive the journey (Wadiwel 2023, 149). These costs are built into the model.

Cape York graziers are aware that there is some controversy around the live export of animals. In 2011, an exposé by the Australian Broadcasting Corporation program *Four Corners* titled "A Bloody Business" (2011) aired footage of Australian cattle being slaughtered in Indonesia that raised animal welfare concerns and led to a temporary ban on the live export of livestock to Indonesia. The ban was short-lived, lasting from June 8, 2011, to July 6, 2011, at which time the government introduced a new export permit system with additional safeguards to ensure the good treatment of animals (Petrie 2019; The Guardian 2011). Despite lasting only a handful of weeks, the live export ban had a significant impact on some of the Cape York graziers. Graziers Bill and Diane told me that, although they weren't personally affected by the live export ban, having sold their cattle just prior to it taking effect, they knew that it had "hurt a lot of people." Alan and Bev told me that they had a particularly tough year in 2011 as a direct result of the ban. At the time of the ban, they had had a substantial number of steers ready to go on a boat, but then they weren't able to send them and had to figure out what to do with the excess stock. As a result, they no longer sell the majority of their cattle to live export, preferring instead to send them to Central Queensland for the domestic market. Pam and Mike were also affected.

As Pam told me, when live export was banned, they were "stuck with the biggest mob of cattle for nearly two years before we caught up with that. And they're eating us out and you can't sell them."

Like others I spoke to, Alan strongly felt that the issues in the industry were blown out of proportion by the *Four Corners* exposé. He said that there were just "three or so bits of shocking footage" that were repeated, and that he believed it was probably just "one shoddy operator" and that most other operators treated their stock better. Alan expressed a point of view that I heard over and over again from various graziers: that if the cattle in Indonesia weren't coming from Australia they would be coming from somewhere else, and these other places may not have the same regulatory safeguards to protect animal welfare during transportation as Australia does. Simultaneously engaging in the live export industry, which even the most defensive of graziers will admit has at least "one shoddy operator," and seeking to minimize their own culpability in the harm that their livestock may experience on the boats and in the distant abattoirs where they end up, graziers' care for their cattle is shot through with these layers of complicity. Insisting that at least *part* of the live export industry is ethical, or involves care, if Australian graziers continue to supply cattle to Indonesia, graziers frame their ongoing engagement in the industry as an acceptable compromise.

Rumors abounded during my time in Cape York about whether or not the footage that *Four Corners* aired was paid for by animal rights groups, or staged, as a way for graziers to further distance themselves from any inhumane animal treatment. But in a situation in which continuing to export cattle to Indonesia is—for most graziers—the only viable option, the care that graziers enact toward their cattle rubs up against the economic contingency of running cattle in Cape York.

There is, in some ways, a sense of decline in the cattle industry in Cape York. Grazier Alan complained to me that cattle trials are no longer conducted in the region. His wife, Bev, elaborated, saying that it has become increasingly difficult to secure loans from banks. She told me that, "the banks will tell you that they don't want to lend to Cape York, because it's no security. They don't see a long-term future." Similarly, Pam told me that,

It's really poor country. No one up here that wanted to make money. . . .

You'd never live here. If you were, you know, wanting to go up in the world or get money, you wouldn't come to Cape York. There's definitely not two bob here. We probably paid tax for about two or three years in our life, we made enough money to pay tax. It's not money making, it's money grabbing and you just live from year to year. And we've got no superannuation or pension or anything. And we can't get a pension because it [their cattle station] is valued probably too high. But it's not. . . . No one that lives in Cape York that stops here . . . Money's not their thing.

However, among the changes to the industry, evidence of inhumane treatment on the boats, economic precarity, and uncertain futures, the graziers I worked with in Cape York remain committed to cattle, and to the place. They endure, living year to year. They make do. However, in their pursuit of a valued way of life, are graziers being uncaring to cattle? Graziers seek to minimize their own culpability related to harming cattle—they insist that live export has been given a media beat up, that Australian beef is more ethical than other options, that the violence they enact toward their cattle in day-to-day management is necessary. For many graziers, though, leaving Cape York and the industry is not an option. As Alan said to me, "that's all I know. Horses and cattle, that's part of our job. Things have changed a bit, recent years . . . but horses and cattle, that's all I ever know."

Domesticating Cape York

The notion that domestication can entail "complete mastery" of humans over animals is false; interspecies relations will always entail unintended and surprising consequences (Lien, Swanson, and Ween 2018; Anderson 2004). As Lien, Swanson, and Ween write, "even in seemingly enclosed spaces, domestication includes complex boundary work, unexpected intimacies, ontological uncertainties, and bodily coconstitution" (2018, 20). While the relation between cattle and humans is profoundly unequal, cattle still retain some agency and can exact their own impacts—unrelated to the human intentions for them—on other species, on landscapes, and on human relationships.

In thinking through the relationship between people and cattle in Cape York, it is important to contextualize the circumstances that brought cattle to the region in the first place. In the wake of the short-lived Palmer River gold rush[3] in the nineteenth century, settlers moved into the region to take up pastoral leases and run cattle. In the introduction to this book, I briefly described the frontier violence of the region at this time, and the devastating impacts on Aboriginal peoples who—despite mounting an effective resistance against settler incursion for many years—were murdered, massacred, and forcibly removed from their homelands at alarming rates. Many scholars have written detailed accounts of this period of frontier warfare (see Loos 1982; Cole 2004; May 1994; Rigsby 1981; Neale 2017), and I will not attempt to do so here. Instead, I want to draw attention to the role of cattle in structuring these early relationships between Aboriginal people and the European invaders.

Cattle and horses enabled Europeans to take over larger swathes of land than would have been possible with human bodies alone, allowing landscapes to be colonized with very few human settlers (Rose 2004, 86). Deborah Bird Rose has described cattle and horses as "the non-human members of conquering societies" (2004, 85). In Cape York, as in other parts of Australia, cattle and the people who owned them worked to dispossess Aboriginal people of their land and resources. The tracts of land most attractive to graziers and livestock were, unsurprisingly, those that Aboriginal people most relied on and frequented; those with reliable freshwater sources. In addition to enabling human conquest, Rose suggests that "cattle become agents of colonization in their own right as they impact on the ecologies they encounter" (2004, 85). Like other ungulates, cattle's hooves compact the soil, altering and damaging sensitive vegetation and ecosystems.

Cattle have played a role in colonization in other places, too. Both Spanish and English imperial forces used livestock as a key strategy for asserting control over new territories in the Americas. Cattle were able to advance into difficult geographies and terrains, sometimes in "advance of empire" (Ficek 2019, S260). As environmental historian Fischer writes, "cattle multiplied even when colonists did not" (2015, 5–6). The English imperial forces considered livestock to be a vital component for transforming Indigenous lands and landscapes into productive land, and for transforming land into private property (Fischer 2015; Anderson 2004; Ficek 2019). Yet, Indigenous people responded to and engaged with these new species in various ways. While cattle transformed both land relations and the makeup of ecosystems themselves, cattle also, often, became an economic resource that Indigenous peoples were able to exploit (Fischer 2015).

In Cape York, Aboriginal people were initially violently excluded by graziers from their newly defined selections. When Aboriginal people hunted or killed cattle or horses, they were confronted with lethal retribution from paranoid settlers. It was not until Aboriginal people were "brought in" to work on cattle stations as a kind of indentured labor that the relationship between Aboriginal people and settlers, as well as between Aboriginal people and cattle, began to shift. In her work on cattle ranching and the enclosure of the commons in lowland Q'eqchi' territory in Latin America, Grandia (2012) has sought to understand why Q'eqchi' people—who, similarly to Aboriginal people in Cape York, were dispossessed of their land and resource base by ranchers—identify with cattle. She describes how Q'eqchi' people often desire to own cattle, dress in the style of cowboys, and have adapted traditional ceremonies and knowledge to use in their interactions with cattle. Grandia writes that this situation is "the essence of hegemony—that people learn to imitate their oppressors and participate in their own domination" (2012, 163). For Aboriginal people in Cape York, though, I suggest that their embrace of a "cattle culture" is not necessarily a mimicry of power. Instead, the ways in which many Cape York Aboriginal people have come to value cattle and cattle work emerges not just from an experience of oppression, but demonstrates the agency and adaptive capacity of Aboriginal people. Cattle, then, emerge simultaneously as both "agents of colonization and objects of desire" (Bonifacio 2023, 9).

Aboriginal People and Cattle Stations

Late in the dry season on a stiflingly hot afternoon, I sat with Lama Lama elder Charlie on the dusty concrete veranda outside the old homestead at Silver Plains Station. Aged in his early seventies with impressive silver sideburns, Charlie enjoyed relating to me stories of his various working experiences, as a stockman in his youth, later as an employee of the Cook Shire Council, and, finally, as a worker in the silica mines on the coast where he worked until he retired. Like many elderly Aboriginal people in Cape York, Charlie lived a mobile existence, sometimes staying at his house in town but more often accompanying his children to wherever they were working or living. Charlie's son, Kevin, was the chair of the Lama Lama Land Trust, and Charlie often came along when Kevin was supervising ranger activities. Charlie would sit in the "old house," observing the rangers and offering stories to whoever of the young rangers stopped by for a chat.

On this day, we were drinking tea and waiting for the relief of the afternoon breeze to come in off the ocean, across the salt pans and down the old airstrip that we faced. Charlie eyed the carloads of young Lama Lama rangers kicking up dust as they drove back and forth from a fence line where they were carrying out repairs. Charlie was reminiscing about his working days. He spoke about fencing when he was a young stockman working on the stations and how they had none of the star pickets or posthole diggers that the young Lama Lama rangers were using. Charlie told me that he would work through the heat of the day building fences, cutting timber using crosscut saws with big teeth that required a person on either end. Fencing was a lot harder back then, Charlie reflected, "no posthole digger, just shovel and crowbar, digging from dawn til dusk." He added that on the days they were mustering the working hours were even longer. Some nights he wouldn't finish until well after dark. Charlie and his family members worked on all the stations around Silver Plains and the nearby town of Coen. All his children were born here, too. "It was all Aboriginal stockmen in Cape York back then," Charlie reflected.

He recalled how they used to muster the cattle from different stations and move them all the way from Coen to Mareeba, a distance of close to 500 kilometers (around 310 miles). Charlie explained that this trip took around three weeks as all the mustering was done on horseback, and they had to be careful not to move the cattle too quickly so that they wouldn't lose condition. The packhorses and the cook would go ahead to set up camp while the ringers would slowly move the cattle, not going too far in one day. Smiling, Charlie told me that at night someone would have to keep watch, riding around the herd and singing to the cattle to put them to sleep. I asked Charlie what he would sing. "Oh, any country and western song that came into your head!" he replied. By the time they arrived in Mareeba, Charlie said, "the cattle were so quiet you could roll your swag out beside them." Winding up his story for the afternoon,

Charlie told me that "it was a hard life, but a good life."

I had various versions of this conversation with a number of older Aboriginal people in Cape York. Shortly after speaking with Charlie, I visited the community justice office in the Aboriginal community of Hopevale to see Bruce and Josephine, an elderly couple involved in the Rinyirru Aboriginal Corporation. I often stayed with Bruce and Josephine when I visited Hopevale, occupying the spare room in their house, and witnessing the flow of children, grandchildren, siblings, and other relatives and kin coming and going, seeking a meal or a chat. Both Bruce and Josephine had grown up in Hopevale and had ancestral and historic connections to various regions across Cape York. Like many people of their generation, the domestic and stockwork of their youth had given way to community leadership roles, and both had diverse working histories, including stints of time with the Aboriginal and Torres Strait Islander Commission (ATSIC), other government agencies, the community-controlled health service, and various Native Title corporations. Both were involved in multiple Native Title cases and had advised during the Royal Commission into Aboriginal Deaths in Custody. At this time, Josephine worked at the community justice office as a cultural adviser to the courts. We were drinking tea, heavily sweetened to cover up the chlorine-heavy taste of Hopevale water, and a half-eaten packet of store-bought biscuits sat open on the plastic table. The aged air-conditioning unit whined persistently in the background. Bruce, having heard I was interested in the time he spent as a ringer in his youth, had stopped by Josephine's office on his morning break from the local Aboriginal-controlled health service, Apunipinya, where he worked as a community counselor.

Bruce launched enthusiastically into a series of stories from his mustering days, mostly discussing memorable incidents involving horses. He told me that he had gained a reputation for quietening down spirited horses and was often called out to different stations to lend a hand. Bruce recalled how at the start of each mustering season the head stockman would pick the stockmen he would employ. "A bit like a football team," Bruce smiled, adding that teams of stockmen were generally composed of people from the same clan or extended family. I asked Josephine if she worked on stations as well. Laughing softly, she told me that on one occasion she remained "in the saddle" until eight months into a pregnancy.

"It was a hard life but a good life," Bruce told me, his words echoing Charlie's a few weeks earlier. "Working on the stations. Dawn 'til dusk, in the saddle."

For many Aboriginal people in Cape York, grazing continues to be spoken about as a legitimate use of land, and a legitimate industry for Aboriginal people to be engaged in. Many Cape York Aboriginal people who have worked in the cattle industry consider cattle to belong in the region in some sense and see cattle work as an appropriate vocation for Aboriginal people, a situation common to other contexts in Australia (Smith 2003b; McGrath 1987; Gill and Paterson 2007). In Cape York and elsewhere, Aboriginal engagement in the pastoral industry has resulted in Aboriginal relationships to land becoming "deeply interculturally entangled" (Ottosson 2012, 191–92). Many of the people I spoke with who were former stock workers described coming to know their ancestral homelands intimately through the processes of traversing the landscape during mustering with older Aboriginal men. Anthropologist Benjamin Smith has described the convergence of cattle management techniques with traditional ways of relating to and coming to know the Country as a kind of "syncretic interpenetration" (Smith 2003b, 32). Importantly, engagement in the pastoral industry allowed many Aboriginal people to continue to live and work on their areas of traditional connection, enabling them to continue to enact care toward the land and ancestral spirits that dwell there through the use of fire and ritual activity (Smith 2003b; 2003a).

It is well documented that engagement in the pastoral industry allowed Aboriginal people in Queensland to escape some of the more restrictive measures enforced by the Aboriginals Protection and Restriction of the Sale of Opium Act 1897,[4] referred to widely simply as "the act," and to avoid removal to centralized missions or reserves, while maintaining cultural connections to significant landscapes through continuing ritual action and giving birth to children on the Country (Gill and Paterson 2007, 127; May 1994; Ottosson 2012; Strang 1997, 29, 33). Indeed, this is the case for at least some of the Aboriginal stockmen and women I worked with. One older Lama Lama man, Percy, told me that it was only by virtue of his parents' employment at Silver Plains Station that he was able to avoid the removal from Yintjingga to Bamaga in the 1960s, which affected much of the Yintjingga Lama Lama community.[5]

Percy was able to continue living on his ancestral homelands through his parents' work, and then his own employment, in the cattle industry. It is important to note that Aboriginal women comprised a significant part of the workforce on stations in Queensland, as both domestic workers and, as Josephine's experience demonstrates, as stockwomen (May 1994, 51; Simone 2016).

Many Aboriginal station workers received little or no pay during the period of substantive Aboriginal employment in the pastoral industry. Graziers Mike and Pam, for instance, employed an Aboriginal couple, Betty and Winston, for many years. Betty received no wage, but her extended family was fed and housed at the station. To Pam, this was a payment "in kind" that was an accepted practice at the time on cattle stations. Pam recalled how a number of Betty and Winston's extended family camped out by the old quarters down the bottom of the garden where Mike and Pam had lived before they built their current house. Betty and Winston's daughter Daisy, too, discussed how Winston worked from the age of thirteen on stations and was not always paid wages for his work. Daisy did not attribute any blame to Mike and Pam for this arrangement and told me she was "pretty sure" the family paid Winston a little bit, an arrangement confirmed by Pam. This hesitance to attribute blame for un- and underpaid wages to White graziers was common among many older Aboriginal stock workers and is a situation that is echoed in other contexts in Australia.[6] Several people discussed the injustice of their unpaid wages with me but were eager to deflect the blame from graziers to particular police officers, protectors, or the Queensland government in general.

People worked alongside each other on cattle stations in Cape York and formed relationships, although their work was never on equal terms. However, this was rarely discussed by older Aboriginal people. Despite instances of violence, racism, and cruelty, despite the exploitation of Aboriginal labor for little or no material gain for the Aboriginal stock workers, and despite the legacy of stolen wages in the grazing industry, older Aboriginal people like Charlie, Percy, and Bruce tend to frame their experiences of stock work as largely positive because of the animal husbandry skills they gained, their ability to remain living and working on their ancestral homelands, and because of the relative independence that their wages, though low, allowed them. The "hard but good life" that such work entailed is held up against the under- and unemployment of Aboriginal people from the 1960s onward, as Aboriginal people's ability to find work was affected by legislative changes like the introduction of Equal Wages and the reorganization of the rural industries toward increased mechanization and smaller workforces.

Asymmetries and Change in the Pastoral Industry

Equal pay was introduced in Queensland in 1966 (May 1994), yet many White grazing families continued to employ and underpay Aboriginal stock workers and have small communities of Aboriginal people living on their stations into the 1970s, owing to lax government oversight. Around the same time, grazing and agricultural industries around the world were rationalizing and mechanizing their practices in what has been called the "Fourth Industrial Revolution" (Gibson 2019, 140). In Cape York, this revolution translated to increased reliance on fencing, trap paddocks, helicopter mustering, the use of nutritional supplements called "lick," and the use of four wheelers instead of horses. These changes occurred gradually, with fences often being the first and one of the most significant changes. Each of these transformations enabled graziers to rely on smaller and smaller workforces, until the situation today where many stations in Cape York are run by a husband-and-wife team, with occasional assistance from grown-up children, neighbors, and contractors. As stations employed fewer workers, many Aboriginal people shifted into towns like Coen and Laura. Without the domain of common work, Aboriginal people and White grazier families no longer had the same level of everyday, incidental interaction that they had previously shared. Across cultural alterity, these shared work experiences—even when occurring in the context of asymmetrical power relationships—provided the grounds for intercultural relationality that nowadays is absent, particularly among the younger generations of Cape York residents.

There is a sense among some the graziers I worked with that things were better "back then," in terms of the way of life and sense of community that using horses and large workforces entailed, and because of the close relationships that existed between Aboriginal and non-Aboriginal people. The golden era of "back then" echoes in diverse rural settings around the world, where people hark back to "a dreamtime before things got out of whack" (Lepselter 2016, 88). The "before" time, in Cape York, in the American West, and elsewhere around the world, was—in the eyes of the graziers—ruptured by government intervention. Importantly, graziers frame things as better before the lifting of the act and the introduction of equal wages for Aboriginal people. Clearly, generations of these grazing families benefited greatly from Aboriginal people as both an inexpensive labor force and people highly skilled in stock work and possessing detailed knowledge of the land. Without this labor force, cattle stations have become less financially viable, leading to many grazing families selling their leases and moving elsewhere. Graziers along with some older Aboriginal stock workers tend to read the situation of Aboriginal engagement in the grazing industry as mutually beneficial, rather than an exploitative relationship.

Today, many blocks of Aboriginal freehold administered by land trusts are used for grazing, and the Indigenous Land Corporation has established several cattle enterprises in Cape York with the aim of training young Aboriginal people with an interest in stock work. As well as engaging in the cattle economy, Aboriginal people frequently compete in and attend rodeos and bull rides in Cape York. These are considered important social events, and the prowess of particular individuals is discussed with pride. Across cattle stations, freehold land, and national parks, the sight of cattle wandering is a common one. Traveling with Lama Lama rangers throughout their Country, people would often point out cattle in the scrub and refer to them as *minya*—a word that means an animal for eating, that is also commonly used to refer to wild pigs. Cattle and feral pigs, while introduced, are considered important bush foods and, particularly for the Lama Lama rangers living on their homelands, comprise a large part of their diet.

While cattle are always understood to be an introduced species, they now occupy an intractable and valuable space in the way that many Cape York locals imagine the character of the region. A similar relationship between Aboriginal people and cattle exists elsewhere in Australia where Aboriginal people have historically worked in the grazing industry. This "cattle identity" (Baker 1999) has influenced how people read and describe the bush. The intercultural space of the multiethnic pastoral industry has also shaped the kinds of environmental knowledges that settler-descended graziers possess.

Environmental Knowledge

Toward the end of the dry season, I clambered into a battered Toyota with cracked vinyl seats and no windows with grazier Bill. Bill and his wife, Diane, in their sixties, live on a lease in central Cape York. Bill lives and breathes cattle work and is deeply pragmatic. With a wry smile, he once told me that he's had good dogs and good horses, but he's never had a pet. Unlike some of the women in other grazing families, Diane prefers to stay close to the homestead and doesn't get too involved in the cattle work. She often spoke longingly to me of going traveling, visiting the kids and grandchildren who lived on the west coast of the Cape and farther south. Some years prior, Bill and Diane had won a camping trailer in a fishing competition. It sat—pristine and unused—in the shed beside the house. For Bill, the idea of traveling was a little uncomfortable, a little unnatural. He said that he didn't like to be far from home for long.

Warning me that the ride was likely to be dusty, Diane sent me off with a scratched pair of wraparound fuel station sunglasses to protect my eyes. In the cargo rack of the vehicle were large bags of a mix of nutrients and salt for the cattle, widely referred to as "lick." We were going on a "lick run" to deposit these huge dusty bags of supplements into drums in makeshift sheds at various points around Bill's lease. At each shed that we stopped at, Bill and I heaved the bags out of the back of the Toyota, slicing open the tops with a small, serrated knife, and pouring the contents into the drums. The cattle were familiar with the sound of the truck and came trotting over eagerly. As we drove across the property, through lightly wooded forests and across dry creek beds, Bill pointed out different features in the landscape. Across a hillside that to my eyes appeared more or less homogeneous, Bill distinguished four or five native grasses, some of which had died back in the dry. He described to me how they looked at different points in the year and spoke about which are the most beneficial to cattle. Shortly afterward, we trundled along a dirt track atop a rocky ridge and Bill told me how years of trial and error had taught him to position the various tracks running through the property to minimize erosion during a big wet season. He told me about mistakes he made in his early days on this station—putting roads in the wrong places, trying to run too many head of cattle.

Lick runs, like the one I went on with Bill, occur increasingly frequently through the dry season as the country dries out. They provide an opportunity for graziers to keep an eye on the welfare of their cattle and the landscape, noting where water and grasses are more and less plentiful. The nutrients in lick help to improve the condition of cattle, and toward the end of the dry season are important for providing cattle with an enzyme to help them digest drier and less nutritional grasses, frequently the only grasses left growing by this time. The introduction of lick transformed the cattle industry, and—for a period—allowed properties to be overstocked as graziers were no longer limited by the supply of high quality, nutritious grasses. Most experienced graziers today acknowledge that they have gone through periods of what Bill has referred to as "flogging the country" by pushing the land beyond its carrying capacity. Seeing that this is an unsustainable practice that leads to a whole host of other environmental issues, most experienced graziers have subsequently determined how many head of cattle their land can comfortably support and tend not to go beyond this.

Bill's detailed environmental knowledge and reflections on his management practices are characteristic of the graziers I spent time with in Cape York. Observing changes in the landscape, weather patterns, and impacts of cattle is, for graziers, a key practice that substantiates their sense of belonging to land, and their confidence in their abilities as land managers. For graziers, this environmental knowledge pertains to and is shaped by their relationship with cattle.

Many Cape York graziers confidently assert that they know their land better than anyone else. In many cases this assurance is grounded in much deeper attachments than a familiarity based on long-term experience. Their assertions about belonging instead refer to the multigenerational ties that graziers' families have had to specific land areas and a combination of knowledge passed down and gained through experience of physical work across their land over a lifetime. Settler-descended rural people come to have a sense of belonging to land through consistent, long-term, and intimate interactions with the land, developing experiential embodied knowledge of the landscape through years of manual labor and firsthand experience (Dominy 2001; Gill 1997; Ottosson 2016; Strang 2004b, 2004a, 2014). It is through using places—traversing them by foot or on horses and quad bikes, returning to particular places that are good vantage points to survey stock, and naming places—that places become meaningful (Gray 1999; De Rijke 2012). Labor, time, and embodied experience emerge as central to how graziers relate to land.

Graziers move across their leases regularly, maintaining familiarity with their stations, which provides the context for ongoing environmental observations and, thus, reaffirms their sense of belonging to the land. This belonging is achieved through a kind of Lockean mixing of labor and land to "grow native roots" (De Rijke 2012) and be reborn as a "[son or daughter] of the soil" (Jackson 2006, 98; Garbutt 2011, 187–88), a key element of White pioneer autochthony and belonging (Geschiere 2009). In Cape York, as elsewhere in Australia (see Gill 1997; 2005), graziers understand their memories of past work, achievements, and failures to be socializing the land. The result is a "slow thickening of meaning" as interactions over time gradually socialize the landscape (Williams cited in Strang 2004a, 38); these are what Strang discusses as cultural landscapes (2004b). Rural people place significance on laboring on the land, experiential knowledge, and obtaining an "embodied sense of place" (Ottosson 2016).

This kind of intimate, embodied knowledge of the land bears similarities to the knowledges that are highly valued by Aboriginal Traditional Owners who, in addition, assert such environmental knowledge as a kind cultural marker (Dombrowski 2001). For many Cape York Aboriginal people, such embodied environmental knowledge is now achieved through ranger work, either for Queensland Parks or an Aboriginal ranger group. On one occasion, I accompanied Lama Lama senior ranger and cultural officer Patricia on a mangrove survey on a clear and crisp day in the early dry season. Patricia, small, energetic, and incredibly knowledgeable, takes her work as a land manager and cultural teacher for younger generations very seriously. She has a soft voice that you have to pay careful attention to in order to hear. Probably owing to her personal interest in anthropology (I was not the first anthropologist she had worked with), she was generous with her time and insights, often explaining to me how particular people in the ranger team were related, or what a phrase or term meant.

Along with a handful of junior rangers and a visiting ecologist, we traipsed across salt pans, photographing and collecting samples of dif-

A Lama Lama ranger talking about plants, 2018.

Fresh growth, Hillview Station, 2018.

ferent mangrove leaves and flowers. As we walked, the soft salt crust cracking under our boots, Patricia frequently made the group pause in order to show me various bush foods: sugarbag honey, small sweet berries referred to half-jokingly as "dog's balls" because of their shape, and green ants. Patricia asked me if I had eaten green ants yet, and when I said that I had not, she deftly caught one and plucked off its abdomen for me to eat. I tasted the small green jewel of the ant abdomen obligingly. It had a tart and refreshing flavor. "You should eat these when you are sick with a cold or flu," Patricia informed me.

She told me that she started going out and "walking around the Country" with senior relatives when she was a toddler. She explained that when she was small, it was easier for her to learn the language words for plants, animals, and places. Before she reached school age, Patricia and her siblings lived with their parents and older kin on the cattle station that the adults were employed at. It was in this context that she walked the Country with her relatives, gaining the embodied and experiential environmental knowledge that she now works to pass on to younger generations through her ranger work.

This kind of connection to and care for land was widely documented by twentieth-century anthropologists working in Cape York.[7] In the ethnographic record, and today, land is socialized through physical presence, imbued with memories and stories, and afforded ancestral sentience. The environmental knowledge that people like Patricia possess is both comprehensive and situated, particular to specific places, and emerging from time spent on Country with other knowledge holders observing and taking part. The significance of physical engagement in maintaining a mutually constitutive relationship between land and Aboriginal people is common around Australia (Munn 1970; Myers 1986, 2002; Merlan 1998; Povinelli 2016). Speaking of the people who live in Belyuen in the Northern Territory, Elizabeth Povinelli suggests that connection to land is based on flows of substance and materiality, as land and people coconstitute each other: "various kinds of activities produce various kinds of substances. Hunting, ritual, birthing, burying, and singing produce *language, sweat, and blood, urine, and other forms of secretions*—with each activity having its own embodied and rhetorical intensities and intensifications" (2016, 78, emphasis in original). To Povinelli (2016), then, Aboriginal relationships to land are constituted by a mutually embodied obligation in which land and people are coconstituted.

As well as the transfer of substances, the significance of walking around Country, spending time, and becoming attuned is deeply significant. For many Aboriginal Traditional Owners, the process of learning about the land is lifelong and occurs in a variety of formal and informal contexts. For Patricia, her education began during childhood, when she and her siblings lived on stations with their parents. Other Traditional Owners spoke to me, variously, of the significance of station work, of mapping trips with elders during the early days of land claims, and of ranger work for learning environmental knowledge and gaining embodied familiarity of the land. The late Deborah Bird Rose (1992) asserted that educating future generations is a key responsibility for Aboriginal Traditional Owners, writing that "the most informal trips in the bush always involve teaching. . . . Although informal, this form of education is serious; adults impress upon their children that they must learn so that they can take over when the adults die" (Rose 1992, 107). Importantly, Rose points out that even though an individual may be born with rights and responsibilities for land, it is up to the individual to "develop their own relationships with country" (1992, 108) through becoming familiar with the tangible and intangible aspects of landscapes.

For many Aboriginal Traditional Owners today, such a mutually embodied obligation with land has been constrained by land tenure arrangements and complex demands on people's time, which restrict people's abilities to interact with the land. The generation that came of age in the wake of Equal Wages and the reorganization of the rural industries had less opportunity to live and work on their homelands. The Aboriginal ranger profession, today, has changed this, and those Aboriginal Traditional Owners who are employed as rangers are now able to maintain an ongoing physical connection with the land. For older generations of Lama Lama and Kuku Thaypan people, this connection, maintained through ongoing physical relationships with the landscape, was enabled by engaging in work on cattle stations.

———

Aboriginal engagement in the pastoral industry was, initially, the result of their being violently dispossessed of their land and resource base by settlers in the nineteenth and early twentieth centuries. However, over time Aboriginal people have come to value cattle and the industry, with cattle work coming to be seen as a "proper Aboriginal pursuit" (Smith 2003b, 33). Aboriginal environmental knowledge has been transformed as a result of the "syncretic interpenetration"(Smith 2003b, 32) of grazing with precolonial forms of environmental knowledge. Settler-descended cattle graziers, too, have incorporated particular kinds of Aboriginal environmental knowledge into their management practices and broader sense of the landscape. This is particularly evident in terms of identifying and harvesting bush foods, but also is apparent in the tracking skills that many older graziers employ and in fire management.

Cattle, and the industry that structures their continued existence in the region, have emerged as a kind of common ground for Aboriginal people and settler-descended graziers. While patterns in work and working relationships have shifted over recent decades as a result of legislative changes and various developments in the rural industries, the era of multiethnic engagement in the pastoral industry is discussed with regularity and fondness by older generations of Aboriginal and settler-descended people alike. There is no doubt that the historical pastoral industry was inherently asymmetrical; while settler-descended graziers were working to establish stations and businesses that could be passed onto descending generations, Aboriginal stock workers received meager pay—if any. Yet, the intercultural zone of encounter that the pastoral industry fostered has resulted in some shared environmental knowledges and values that continue to hold relevance for many people.

Cattle themselves represent multiple and competing values for different people. While in some ways functioning as agents of colonization, cattle are understood to contingently belong in the region by Aboriginal and settler-descended peoples alike. Cattle are valued for their instrumental use. They are a source of food and livelihood for many in the region. However, taking into account the ways in which graziers "pay attention" to cattle and their needs, and enact care in their work with cattle demonstrates an attunement to what cattle experience, which goes beyond seeing the beasts as simply serving human purposes. As van

Dooren (2014; 2019) and Haraway (2008) have noted, human-animal relations can simultaneously be relations of violence and of care.

For cattle graziers and some Aboriginal Traditional Owners, it is considered possible to simultaneously have a profitable cattle business and effectively manage the land. However, the notion that cattle "belong" in Cape York is contested by Queensland Parks and other conservation groups. To these groups, cattle represent degradation of the environment. As a result, cattle are framed as a pest species that need to be excluded from national parks and sensitive ecosystems. Such a position is locally contentious, provoking questions about working landscapes, conservation, and control that I turn to in the next chapter.

Cattlemen at a rodeo, 2019.

TWO

Fences

After dinner one night at Hillview Station, grazier Pam and I sat chatting idly in the living room under the bright fluorescent light, half watching the day's news on the television. Pam's husband, Mike, was on the phone in the adjacent office. We could hear snippets of his conversation, and it was evident that he was becoming agitated. After some time, Mike joined us. He had been speaking with his nephew, grazier Alan, who lived on a cattle station not far from Mike and Pam's place. Like Mike and Pam, Alan's pastoral lease shares a boundary with Rinyirru National Park. Like Mike and Pam, Alan's cattle had wandered through broken fences and washed-away floodgates into the national park the wet season prior and he was eager to obtain the necessary permit to muster his cattle back. According to Mike, Alan had been speaking to the Parks office about the permit and was experiencing difficulty with staff members he perceived to be unhelpful—a situation that was making him "wild" (angry).

Mike told me that he and Pam had lodged their request for a permit to muster the park around seven weeks prior. They had been told that the process would take three weeks, but at this stage they still had not received their permit. As Mike discussed this delay, he and Pam became increasingly upset, their complaints broadening out from annoyance at the delay with the mustering permit to anger at the issue of the contested ownership of some of the cattle in the park. "They're our cattle! Our progeny!" Mike said, raising his voice slightly. "Who do they think the cattle belong to?" Pam chimed in, "How can they belong to one person and then just change ownership!"

Frustrated, Pam turned to me and expressed her exasperation at being surrounded by national parks. Not just Rinyirru National Park but others, too. She listed three neighboring parks, all of which Hillview Station shares a boundary fence with. "Our cattle end up on all of these parks and we have to apply for permits to muster them all." As Pam explained to me, the boundary between their station and Rinyirru National Park is along a river, and as such the boundary fences are inevitably damaged each year during the monsoon when the rivers swell. When the rivers come back down again, cattle wander through these broken fences to the better pastures that are located within the bounds of the park. Pam told me that she would need to calm down and stop talking about Parks and their cattle and the mustering permits because she was getting "wild." Yet, both Mike and Pam continued speaking, listing their grievances. They were both staring at the television but not watching, staring but continuing to speak about their frustrations with Parks. Staring, and continuing to explain why they were both "wild."

Tensions over boundary fences and the cattle management strategy have come to characterize the relationship between Queensland Parks rangers and cattle graziers. While graziers may articulate their complaints in terms of infringements on their rights, what the contestations over the cattle management strategy demonstrate are the contrasting ideas people hold around what kind of landscape (and land use) is appropriate in the region. Where Queensland Parks managers seek to protect and preserve a "natural" and "native" landscape, with strong biodiversity values and tourism potential in line with their mandate, graziers and many Aboriginal Traditional Owners are laboring to create what I have been conceptualizing as *workable landscapes*. In this chapter, I show how graziers' attempts to hang on and make do rub up against formal conservation structures. The state government, through the auspices of QPWS, seeks in some ways to wrangle graziers, to reshape them into more compliant neighbors through management agreements and permit systems. However, despite a stated rejection of environmentalism and conservation, many graziers seek to forge workable landscapes and livable lives, caring for landscapes and ecosystems in ways that often run counter to the accepted logic of National Parks.

Boundary Fences, Exclosures, and the Rinyirru National Park Cattle Management Strategy

Cattle have been and continue to be a significant management challenge for the Queensland Parks and Wildlife Service in Cape York. In the largest park in the region, Rinyirru National Park, a locally contentious cattle management strategy has been implemented, drawing the ire of graziers and bringing to the fore the conflicting aspirations of Queensland Parks managers and rangers and some Aboriginal Traditional Owners. Part of the cattle management strategy involves legislative backing for Queensland Parks to begin culling cattle, which are considered a pest species within the park. To understand why the cattle management strategy is such a source of contention, it is important to understand the history of the park itself and to chart the shifts in what Queensland Parks has considered to be permissible over the years.

Before its establishment in 1979, Rinyirru National Park was a cattle station. More precisely, it was two cattle stations: Laura Station to the south, established in 1879, and Lakefield Station to the north, established some years later. Many of the places in the park have retained their names from the station days, and the old homesteads still stand. Both are heritage listed timber houses, raised high on stilts, now in various states of disrepair. The old house from Lakefield Station is fenced off. The Laura homestead itself remains inaccessible, but underneath the house is a quasi-museum, with a few pieces of farming paraphernalia and faded, dust-encrusted laminated photographs accompanied by writing about the pioneer history of the site. The cramped, dark, and tiny sheds where Aboriginal workers were housed still stand and are open to the public. They are made of corrugated iron and have dirt floors; the interiors are hot and stuffy.

Some cattle graziers and Aboriginal Traditional Owners living and working today remember the time before these stations were appropriated and transformed into Lakefield National Park—later renamed Rinyirru National Park when Native Title was recognized over the park and it transitioned to a joint management structure between Queensland Parks and the Rinyirru Aboriginal Corporation. The memory of the park as a cattle station—experienced as well as inherited—remains salient in various people's minds. When I was undertaking fieldwork, most people, including Aboriginal Traditional Owners, continued to refer to the park by its previous name: Lakefield.

Rinyirru National Park now borders other national parks and an Aboriginal Land Trust to the east, but to the north, west, and south the park shares a boundary with various cattle stations. In the 1970s, cattle stations in the region were unfenced, and for several decades after the park was established, neighboring graziers were allowed to continue grazing their cattle on the park. In 2011, Queensland Parks instigated efforts to destock the park, with rangers working alongside the Rinyirru Aboriginal Corporation and other Traditional Owners to conduct large-scale musters in multiple years (Queensland Government 2013). According to Ray, the ranger in charge for Rinyirru National Park, the first boundary fences were erected in the 1980s, with more recent fences put up in 2007. As of 2013, around 350 km (around 217 miles) of the park's boundary remained unfenced or in need of repair (Queensland Government 2013, 8).

Neighboring graziers point to failures in the efficacy and completeness of these boundary fences. There are certain sections of the boundary that remain unfenced, as at points the boundary crosses wide rivers that flood in the wet season but at other times become dry beds that operate as a kind of thoroughfare for cattle. Aside from these gaps in the fence that exist year-round, the monsoon brings with it flooding and debris, which generally dismantles floodgates and brings down fencing wires and posts. Certain areas of both the park and the neighboring stations remain inaccessible until sometime after the wet season has ended and the floodwaters receded, leaving a relatively long window for cattle to find their way into the park. Graziers must then apply for a permit to muster their cattle back, framed by certain stipulations.

Thus, there remains two sources of cattle in the park today. There are cattle that are wild and descended from left-behind herds from the prefenced days, called feral scrub cattle and also referred to as "Cape York reds." There are also cattle that have more recently found their way into the park through various avenues. These cattle are brahman and are worth substantially more money than the Cape York reds. In some cases, they are the result of careful, generations-long breeding programs. Importantly, these cattle continue to breed and calve within the park boundaries. There are many reasons that cattle continue to enter the park, and in certain cases these involve some level of human intervention or at least complicity. It remains clear to the graziers that the total removal of cattle from Rinyirru National Park is impossible given the nature of the landscape and the weather. Fences just do not stay up, and cattle will inevitably seek better grass and more reliable water sources that may be inside the park boundary. However, for some Parks staff, the ongoing presence of cattle on the park is understood as an indication of intentional wrongdoing on the part of graziers. In attempting to implement a cattle management strategy and remove all cattle from the park, Queensland Parks rangers experience graziers as a kind of lingering thorn in their side, as graziers disrupt and resist the management strategy in various ways.

Grazing ungulates like cattle have a variety of impacts on savanna landscapes, affecting both vegetation and vertebrate communities (Neilly and Schwarzkopf 2019; Arcoverde, Anderson, and Setterfield 2017; Mihailou and Massaro 2021). Through their grazing, cattle impact grasses, resulting in deep-rooted plant species being replaced by shallow-rooted species. Grazing generally leads to perennial grass species being replaced by annuals and can eliminate particular native grasses from areas where grazing pressure is high. Because Australian savannas have historically experienced less grazing pressure than other savannas, such as those in Africa, many native grasses and plant species remain vulnerable and unable to compensate for the impacts of grazing (Mihailou and Massaro 2021). As such, even a relatively short period of intensive grazing can cause irreversible damage to savanna vegetative communities. This has several flow-on effects, including increased erosion and runoff of the topsoil during rainfall, resulting in a "leaky" or dysfunctional landscape, as well as the creation of disturbances that allow invasive plant species to spread and thrive (Ludwig et al. 2001; Mihailou and Massaro 2021). Such impacts are not distributed equally across the landscape, and the damage from grazing tends to be more severe close to water points, which are likely to experience a high volume of animal traffic. The presence of ungulates at water points also affects waterways, as their excrement contaminates the water resulting in an increase in nutrient levels (specifically nitrogen and phosphorous), which can lead to eutrophication and algal blooms (Mihailou and Massaro 2021). In addition, the reduction in herbaceous vegetation due to grazing can mean that fires are smaller and less intense than would otherwise be the case. Lower intensity fires, in turn, contribute to what is known locally as "woody thickening" in the landscape, that is, the proliferation of melaleuca scrub (Mihailou and Massaro 2021).

A 2013 management statement regarding Rinyirru National Park states that, "Branded and unbranded cattle occur on Rinyirru. They degrade wetlands, introduce pest plants, change the structure of the vegetation, and form deep pads along fence lines and in places where they exit watering points" (Queensland Government 2013, 7). There are no data on the actual numbers of cattle remaining in the park today, although an article in a local newspaper estimated that in 2016 there was approximately 1,000 cattle wandering the park, which across the 544,000 ha land area of Rinyirru represents a fairly low density (Geiger 2016a). As such, and as the management statement points out, impacts tend to be concentrated around high-use areas, such as fence lines and water points.

Cattle have negative impacts on the ecosystems of the park; however, Queensland Parks' focus on removing cattle is nonetheless worthy of analysis. While the desire to have the park free of cattle is doubtless related to the conservation objectives of the park, it is also about presenting a particular experience of "wilderness" to tourists. As senior ranger Tamara told me, "I think, ultimately, from a tourist point of view, we used to get a lot of complaints about cows wandering around the campsites . . . it's not a cattle property it's a National Park . . . [and] when they can see them doing the damage . . . it's a feral animal." Similarly, the ranger in charge for Rinyirru National Park told me during an interview that "we are a conservation park, and we shouldn't have cattle on the park. That's just the bottom line. And we're getting there."

Queensland Parks has been trying to implement the cattle-management strategy in Rinyirru National Park since 2013. This strategy was created by Queensland Parks and approved by Rinyirru Aboriginal Corporation and consists of four components. The first step was an initial muster conducted by Queensland Parks and Rinyirru Aboriginal Corporation in-

Mustering camp, Hillview Station, 2018.

Repairing a fence damaged during flooding, Tidewater Station, 2019.

tended to remove as many cattle as possible from the park. The second component was the creation of a mustering permit system that would allow graziers to come and retrieve their stock. The third phase was to maintain boundary fences, and last, the remaining cattle were to be culled.

The quite radical step of going ahead with culling cattle was, at the time of my fieldwork, unique to Rinyirru National Park for Queensland Parks. There is a sense among the on-Park staff at Rinyirru National Park that they are operating as a kind of test case, to see whether similar strategies could be implemented across the state. Under the legislation and the cattle-management strategy, Queensland Parks is entitled to cull all remaining cattle on the park, but this is unlikely to ever eventuate, for a range of complex reasons. Ray told me that while Queensland Parks is entitled to cull branded cattle, it "wouldn't pass the *Courier Mail*[1] test." That is, public outrage means that this is not a viable option. Opinions vary among the Parks staff about the ethics of culling branded stock. While Ray spoke about the cull as unable to go ahead because of the risk of public outrage, another ranger with a history of working on cattle stations said firmly and seriously that "you can't shoot branded cattle. It's illegal." The most senior staff member for the region, Anthony, spoke about the need to maintain a relationship with neighboring graziers. As he told me during an interview, "sometimes it has to be a robust conversation between [Queensland Parks and the neighboring graziers], and say, 'right, we know your cattle are here, get them off the park.'" Anthony told me firmly that only branded stock that had been "relinquished to the State" would ever be culled. The only cattle, then, that are culled are cattle referred to as "cleanskins," which are cattle without a brand.

The cleanskin is a shadowy figure. Brahmans mixed (to varying extents) with Cape York reds, cleanskins are not quite *Bos taurus*, not quite *Bos indicus*. They are considered by some to be feral animals and by others to be a valuable economic resource. Under the legislation, cleanskin cattle are the property of the Traditional Owners who hold Native Title over the park, although the ownership of these cleanskin cattle, and what ought to happen to them, is widely disputed. For graziers, unless they are very obviously the much less valuable Cape York reds, cleanskins are considered to be the unbranded progeny of their branded cattle—their property. To them, these progeny should not be included in the category of cleanskin that belong to the Traditional Owners. They assume that for something to be truly considered a cleanskin it must be further removed from their branded stock and thus a descendant of feral cattle that have lived on the park unchecked for years. Until relatively recently, graziers have been able to maintain this assumption, aided by Queensland Parks' "soft touch" approach. Graziers claim that being able to collect cleanskin cattle during their musters of the park made the exercise more economically viable, too. At the time Queensland Parks was happy to turn a blind eye, given that their ultimate goal was to remove as many cattle as possible from the park. As Queensland Parks move further into their cattle-management strategy, this "soft" approach has shifted into one that is more hard line. From a management perspective, this shift is a logical consequence of moving toward complete removal of cattle from the park. For on-Park staff the change in approach seems to be related to a sense that graziers have been taking advantage of Queensland Parks, requiring a strong message to be sent. These differing understandings of who owns the progeny of branded cattle have only come to a head as Queensland Parks has gone ahead with their plan to cull.

For Aboriginal Traditional Owners, though, cleanskin cattle and the management strategy in general represented an opportunity to build skills and competencies in young aspiring stock workers, and to generate revenue for the Rinyirru Aboriginal Corporation. When the management strategy was first implemented in 2013, there was hope that the cleanskin cattle could be mustered by Traditional Owners who would be employed as contractors. The intention was for Queensland Parks to fund the muster, but for revenue from the sale of the cattle to be absorbed back into Rinyirru Aboriginal Corporation. According to several Queensland Parks rangers with whom I spoke, attempts to do this in the past were unsuccessful and expensive. Because of the expansive nature of the park, the contractors were able to gather only a small number of cattle, and Queensland Parks lost around AUD$300,000 (USD$193,800) as a result of the program. Accordingly, Queensland Parks has moved away from expensive attempts to muster, seeing culling as more effective in terms of cost and outcome.

While this has been agreed on through the joint management process, it remains contentious among Aboriginal Traditional Owners who would prefer to see the cattle mustered and sold. One older Aboriginal man who had spent his youth working on stations described the plan to cull as "a waste of cattle," indicating the value (economic and otherwise) he places on cattle. Rinyirru Aboriginal Corporation ranger, Donna, likewise spoke of the plan to cull as a waste. To her, this was more related to the loss of an opportunity to get young people into work than about the waste of valuable cattle. She explained that getting young people "out of the community and into jobs on Country" was a priority for her. Donna believed that allowing young Aboriginal Traditional Owners to muster the park did not necessarily require a large injection of money. "A lot of the boys in the community already have the gear," she told me, explaining that her son and nephew owned quad bikes and had some experience working on stations.

Decisions to cull feral livestock animals on joint managed parks have—in some instances—revealed a wide gap in understandings about whether or not species belong somewhere, and what value is ascribed to these introduced species. Chris Haynes (2009) details how a dispute over culling horses in Kakadu National Park, that were considered by park rangers to be pest animals and by many Aboriginal Traditional Owners to be "bush pets," escalated into a significant conflict, with some irate Aboriginal Traditional Owners eventually leaving the severed head of a culled horse on a park ranger's doorstep. It took years of careful diplomacy to mend the joint management relationship to a point at which it was functional. The native/introduced binary, so pervasive in environmental management, does not always easily map onto whether Aboriginal people considered species to belong on their Country or not. Cattle, horses, buffalo, camels, and even cane toads have been accepted by Aboriginal people as belonging in the land in some sense—sometimes absorbed into their cosmologies and sometimes becoming a part of peoples' lands and lives in more quotidian ways (Riley 2013; Vaarzon-Moral 2017; Robinson, Smyth, and Whitehead 2005; Seton and Bradley 2004; Gibbs, Atchison, and Macfarlane 2015; Trigger 2008). As chapter one details, many Aboriginal people in Cape York have come to see cattle as belonging in the region and understand cattle work to be a "proper Aboriginal pursuit" (Smith 2003a, 33). It is therefore unsurprising that Aboriginal Traditional Owners would prefer to be mustering cattle off the park themselves, as opposed to allowing valuable stock to be culled via a method that does not provide employment or training for young Aboriginal people.

Changes to the Permit System and
the Perceived Overreach of the State

The progression of the cattle-management strategy has meant that the rules for graziers around mustering the park have changed in recent years. Whereas in the past, graziers received a permit that allowed for flexibility and extensions around the retrieval of their cattle, they must now apply to muster a specific section of the park for a three-week period. Certain other stipulations have recently been introduced: Queensland Parks and Rinyirru Aboriginal Corporation must both be given the opportunity to have a ranger or traditional owner present for the muster, and there is increased surveillance around the taking of cleanskin cattle. This new information was communicated to the graziers via email, a form of communication that many graziers remain uncomfortable with. I found myself translating this email, on more than one occasion printed out and presented to me, for graziers when I visited their stations.

The shift to a more formal communication style between Queensland Parks and graziers has contributed to the social divide and mutual distrust between the two groups. Many graziers assert that their relationship with Queensland Parks has become increasingly strained in recent years and cite both disputes around the cattle management strategy and changes in Queensland Parks personnel as key reasons for this. Graziers Pam and Mike emphasized to me that their relationship with Queensland Parks used to be much more positive and mutually supportive. In the past, rangers and graziers frequently coordinated fire and pest-management programs and occasionally socialized, but in recent years there has been a drift away from these friendly relationships.

Pam spoke about this shift as we sat around her kitchen table drinking tea with her husband Mike. She said, "before, National Parks fellas were all . . . They'd done fieldwork, that sort of stuff, they were a bit more our people. They were always glad, they'd come and see the birds and they'd stop here."

Mike agreed, saying, "a lot of them were station people."

"Most of them," Pam continued, "they were all people that were off the land."

I asked if this meant they had a bit more understanding toward the graziers' perspectives.

"Experience," Mike said. "They got experience."

Nowadays, there are few Queensland Parks rangers with this kind of valued "experience," and most rangers lack substantial commonality with, or sympathy toward, the graziers' situation. This increasing social divide is keenly felt by the graziers who now tend to see Queensland Parks staff as faceless public servants with no significant attachment to Cape York, rather than friends with whom they would socialize on occasion. Park rangers have come to fulfill a role akin to police officers in graziers' lives. Indeed, Anthony, a senior ranger for the Cape York region, told me during an interview that the Queensland Parks and Wildlife Service is the second largest law enforcement agency in Queensland, after the Queensland Police Service. While this is technically true, his desire to highlight this to me during our interview reflected his strong focus on compliance in the parks he administers. Compliance is a significant part of what Queensland Parks do. Multiple graziers complained to me that many park rangers are "more like police officers now" than park rangers from previous years. Indeed, I knew of several senior Parks employees who had worked as police officers prior to their appointments with Parks. In the vast and remote interior of Cape York, park rangers are frequently, for graziers, the most tangible "manifestation of the state," aside from the odd traffic cop who might fine graziers for firearm and seatbelt violations. Laura Ogden has described the game wardens of the Florida everglades who censured illegal alligator hunting as bringing "the state's constitutional authority over nature into the field and ma[king] the abstract of government and law immediate and legible" (2011, 137). The same is true of the situation in Cape York, and it is made particularly obtuse through the kinds of antagonistic relationality that the cattle management strategy has produced. When park rangers enforce the cattle management strategy and rules around mustering permits, they are experienced by graziers as the forcible insertion of the state into their lives. Moreover, they indicate to graziers that the state's imagined future of Cape York is one without cattle, in which graziers no longer have a place.

From the perspective of Queensland Parks, graziers are hostile and resistant to the implementation of the cattle management strategy, as they fail to adapt to the current system. Several Queensland Parks staff relayed to me that they believed neighboring graziers had been using Rinyirru National Park as a kind of "free agistment" for years, intentionally allowing their stock to enter the park and graze on the good quality grass there. When I asked ranger Ray whether he thought neighboring graziers were letting their cattle into the park, he replied, "of course they are. Absolutely. No two ways about it. And a typical example of that is if you go around one of the boundary fences and see a spear trap, what is that telling you? And that's just how it's been." Ray told me that he believed a hard-line approach, and the imminent culling of cleanskin cattle, was the only appropriate course of action.

The graziers, though, largely disputed the accusation that they would intentionally allow their cattle into the park. One grazier, Alan, whose station was wedged between Rinyirru National Park and Lama Lama National Park, acknowledged that not everyone is as scrupulous as he claimed to be. In discussing Queensland Parks' antagonistic attitude toward him and his family, he said, "I mean, I can understand it a little bit because people, some people, were doing the wrong thing. On the park. They weren't there to clean their cattle out."

His wife, Bev, elaborated, saying, "they weren't selling everything they got. They were only selling the tops off them and leaving cattle there, stuff like that . . . And that's what [Queensland Parks] got frustrated with. People not shifting everything that they mustered." As Queensland Parks' staff have grown increasingly confident in their belief that graziers have been taking advantage of their soft-handed approach, graziers' have sensed that antagonism and become increasingly antagonistic in response.

The removal of cattle from Rinyirru National Park can be understood as Queensland Parks firmly staking a claim about what kind of landscape is appropriate and worthy of preservation in Cape York. The tightening of the mustering permit system is perceived by graziers to be a kind of ideological struggle about what kinds of landscapes and land uses are supported in Cape York by the state and the broader public. Where grazing was once an accepted practice within the park, its censure is experienced by graziers as antagonistic and an attack on their livelihoods and way of life. To graziers, the enforcement of the cattle management strategy is entwined with a broader set of transformations in the region, which they perceive as Cape York being "locked up" for conservation purposes.

Part of the tension between graziers and Queensland Parks is related to the divergent relationships that these groups have to cattle. Where graziers and many Aboriginal Traditional Owners see cattle as both an important resource and a species that "belongs" in Cape York regardless of where the cattle are grazing, Queensland Parks rangers understand those cattle within the park boundary to be feral animals requiring exclosure and control. The impacts of cattle on conservation values and the visual amenity of the park (ensuring that tourists encounter the kind of landscape they expect) drive Queensland Parks' project to remove and cull the remaining cattle on the park. The multiple categories cattle inhabit within and outside the park boundary, and different values attached to cattle, underlie the conflict over the cattle management strategy. In Rinyirru National Park, cattle have become "creatures out of place" (Scaramelli 2021, 75).

The removal of animals that are traditionally livestock but have become "feral" from protected areas is not unique to Cape York. Similar attempts to de-stock or remove animals like cattle and horses from protected areas exist around the world, precipitating tensions and—on occasion—resulting in violent confrontations between rural people and park rangers (Scaramelli 2021; Riley 2013). In the wetlands of Türkiye, attempts have been made to remove horses from marshlands so as protect the bird species that their grazing is presumed to threaten. However, as Caterina Scaramelli notes, the limited research that has been carried out on grazing impacts in the marsh revealed that "the effect of grazing is not univocal; it is complex and can produce benefits as well as negative consequences" (2021, 76).

In other instances, such as the controversial feral horse culls in Australia's Kosciuszko National Park, the damage wrought by horses to sensitive ecosystems cannot be denied. In Australia, feral horses are generally referred to as "brumbies," and they occupy a privileged position in the national imaginary.[2] Here, the conflict is not related to contestation over whether or not horses damage the park, but about whether brumbies can be understood to "belong" in the region or not. Members of historical alpine grazing families have been particularly vocal in their resistance to brumby culling. These grazing families had lost access to historical grazing leases in the region some decades prior (Dominy 1997), and, to them, the brumby culls were understood as the latest in a long line of attacks on their "way of life" (Riley 2013; Lamond 2023; Hagis and Gillespie 2021; Farley 2023). Yet, the brumby cull proved to be controversial in the public domain as well, which scholars have attributed to the "iconic" status of brumbies, especially in the alpine region of Kosciuszko National Park (Hagis and Gillespie 2021). Concerns were raised over the morality of culling brumbies and the methods used (aerial shooting), despite these methods being routinely used to cull deer and pigs in the park with no public outcry (Riley 2019; Hagis and Gillespie 2021, 234).

Despite Ray's fears around cattle culling "not passing the *Courier Mail* test," in comparison to the controversy over brumby culling there has been little public outcry over killing cattle in Queensland Parks. There are a couple of local news articles scattered around rurally focused publications that condemn the plan to cull cattle (see Geiger 2016a, 2016b; Coghlan 2017; Richardson 2017), but the culling of cattle has not captured the public's imagination in the same way as that of brumbies. Furthermore, the concerns raised in these articles are related to the economic value of cattle, rather than any claims that cattle are iconic or have a status worthy of protection. Yet for graziers, Aboriginal Traditional Owners, and some natural resource management practitioners, the culling of cattle is considered to be a "waste" that is felt on an emotional level. Many people remember viscerally the heartache of a widespread culling operation in the 1980s that was deemed necessary to stop the spread of bovine tuberculosis (Tweddle and Livingstone 1994; Cousins and Roberts 2001). Marvin, a man of Aboriginal and European heritage who works in pest animal management across Cape York, told me that when he took part in shooting cattle as part of this biosecurity program, "bullocks started chasing [him] at night," haunting his dreams.

Graziers' relationships to cattle are entwined with economic aspects. Cattle comprise a significant part of graziers' livelihoods. However, perhaps more significant is the link between cattle and a valued "'way of life" for graziers. Despite most grazing families having diversified their income streams, taking part in roadbuilding, tourism, truck driving, and contracting alongside their cattle concerns, it is running cattle that is central to their self-identification and sense of belonging in Cape York. Like the graziers of the alpine region around Kosciuszko National Park, Cape York graziers experience the cattle management strategy as an attempt to erode their "way of life," and a reminder of the state's presence in their lives. To graziers, the cattle management strategy is the latest iteration of what they see as a broad project to "lock up" the Cape York region for conservationist and Indigenous interests.

The "Locking Up" of Cape York

I encountered a belief among many graziers, some Aboriginal Traditional Owners, and even some Parks staff, that the government intended to "lock up" Cape York from private industry and development in order to preserve the entire region for conservation and Aboriginal interests. The narrative about Cape York being "locked up" seems to have emerged from a combination of actual events, policy shifts, and rumors and has been widely documented (Slater 2013; Neale 2017; Holmes 2011b). The transfer of a number of pastoral leases into National Parks in recent years has fed into the narrative of "locking up." Grazing families who have remained in Cape York through climatic and market fluctuations have seen neighboring leases sold and bought by Parks, and although barely any of these leases were forcibly acquired, graziers have perceived these land tenure changes as the state government's attempt to push them out of the region. Grazier Mike told me: "I heard it from different people, they're trying to push us out of the Cape. . . . The government is trying to get rid of us, out of the Cape. So, this is the way they're going to do it now. Really force us out. That's what I feel anyway."

There has been no talk of Mike and Pam's station being acquired for conservation purposes, despite the recorded presence of the endangered golden-shouldered parrot on their lease. However, their lease is now almost completely bordered by parks and Aboriginal freehold land, a shift that affects them in terms of the difficulties with contested boundaries and mustering permits, but also on an emotional level. On one occasion when I was staying with Mike and Pam, they had some longtime friends visiting—an elderly woman and her adult daughter—who had lived on a neighboring station before it was sold and transferred into Aboriginal freehold. Together, we went to visit the yards and sheds near the boundary fence between the two parcels of land. We wandered around the disused cattle yards and rusting sheds, as everyone recounted stories of mustering the boundary together in the years before fences were put in. These boundary musters are remembered fondly, described as more of a social gathering than anything else. As we piled back into the car to return to Mike and Pam's house, the younger of the two visiting women said to Mike that it was sad to see the way that a lot of stations have gone now. "Really hard," she said, "to see all the blood, sweat, tears and hard work go to waste," as yards and sheds fall apart, and the land is no longer used for grazing.

While the narrative of Cape York being "locked up" is a pervasive one, touted by most graziers in the region, most people are aware that it is not a case of land being forcibly acquired and dedicated as parks. As grazier Alan said to me of many of the stations that have recently been transferred into parks,

> Parks bought them because they were for sale. The reason they were for sale [is because] people paid too much money for them. . . . [They] got in trouble. Parks were the easy way out. And good money. Because what happened with the Parks [is] they sell to Parks, get their dollars, what they want for it. Plus, they get all the cattle off as well.

Alan, then, acknowledges that selling to Queensland Parks is an attractive option to many grazing families, particularly when their cattle businesses are not lucrative and they are struggling with enormous amounts of debt. These comments speak to the economic reality of Cape York: it is extraordinarily difficult to make money grazing cattle. Many people struggle to make do, and many welcome the lifeline of a generous offer from Queensland Parks. The fact that Cape York does not make for good grazing country means that the cattle industry in the region is, enduringly, marginal. Just as graziers were unable to continue to employ large Aboriginal workforces after the introduction of Equal Wages legislation, many grazing family businesses are unable to weather the fluctuations of the market. The nonviability of the industry in this place echoes through time. As senior ranger for the region, Tamara, stated, "obviously the Department did buy cattle properties, but in the same token the graziers could have bought them and used them themselves and they didn't." Tamara and Alan's comments gesture to the fact that many of the pastoral leases that have sold and transferred into Parks in recent years were on land that was not particularly productive for grazing cattle.

There is, however, consternation over the fact that some properties considered to be the most productive in the region became National Parks in the 1970s. Prior to its designation as a National Park, Lakefield Station was considered to be some of the best grazing country in the region. As Cape York Natural Resource Management (CYNRM) employee and Landcare officer Michael explained to me,

> Compared to the Territory, the National Parks that worked there were sort of up in the catchments, the headwaters, the sort of rougher country. And the better grazing country was still grazing country. . . . Whereas here it seems to be a little bit the opposite. . . . Lakefield was one of the most productive—okay, it's beautiful wetlands and has got high conservation value, but they're saying it was one of the most productive places on the Cape. And it's National Park.

Alongside the actual redefinition of land and transfer of grazing leases into Parks and Aboriginal freehold, Cape York has been a site of significant contestation over land use. Long framed as a struggle between " 'green" and "Black" interests (Holmes 2011b, 2011a; Anderson 1989; Langton 1998), the Wild Rivers controversy of recent decades explicitly brought the disjunct in aspirations and concerns between local Aboriginal people and national conservation groups to the fore (Slater 2013; Neale 2017). While conservation groups wanted to see an exclusion zone around waterways—the so-called Wild Rivers of Cape York—conserved and kept free from development, some Aboriginal community leaders vehemently opposed the legislation. Noel Pearson, prominent Aboriginal public intellectual and former chair of the Cape York Land Council, asserted that the legislation would be a "death by a thousand cuts" for graziers and Aboriginal peoples alike (Neale 2012). By no means was Noel Pearson the voice of a unified community of Aboriginal people in Cape York, and it is important to mention that other Aboriginal community leaders spoke out publicly against Pearson and in support of the act (Neale 2012). However, Pearson's comment points toward some shared concerns around the privileging of environmental interests over the livelihoods of local people. As Lisa Slater (2013, 773–74) has indicated, there is a history of Aboriginal Traditional Owners and graziers sharing a mutual suspicion toward conservation groups in Cape York. Some of this suspicion on the part of Aboriginal people is a result of the basis on which early National Parks in the region were declared. As detailed in the introduction, the then-named Archer Bend National Park (now Oyala Thumotang National Park) was established explicitly to exclude Aboriginal Traditional Owners from gaining title over a large tract of their land. This was a true instance of land being "locked up" for National Parks and Aboriginal people forcibly excluded.

The "locking up" that graziers point toward is largely symbolic, and claims that Cape York graziers are being forced out can be widely discounted by considering the circumstances under which many cattle stations become National Parks and Aboriginal freehold. However, graziers are nonetheless experiencing what Dominy has called "a process of rural peripheralization" (1997, 241). What is significant is that the transfer of significant amounts of land into National Parks and Aboriginal freehold is perceived and experienced by graziers as a simultaneously symbolic and tangible attack on their way of life. A vision of a certain kind of nature, landscape, and working relationship to land (Watt 2017; Ogden 2011) is understood as being imposed by Queensland Parks, with the backing of policy, legislation, the state, and the broader (urban) public from "down south." In imposing new rules and regulations around getting cattle off the park, Queensland Parks—to graziers—become part of a vague coalition of "undifferentiated power" (Lepselter 2016, 103) composed of environmentalists, a distant urban-dwelling public, and the government that graziers experience as eroding their autonomy. Surrounded on all sides by parks and Aboriginal freehold and struggling through an unfamiliar bureaucratic system to apply to muster their cattle back from parks, graziers tend to position their livelihoods and way of life as being under threat.

Working Landscapes

On a humid and overcast day at the rodeo grounds in the town of Laura, I attended the annual grazing forum with Pam and her family. The forum is a two-day workshop, hosted by Cape York Natural Resource Management (CYNRM) and agricultural industry body AgForce, comprising short educational or engagement sessions run by a variety of groups including the Meat and Livestock Association, the Queensland Police Force, Border Force,[3] an all-terrain vehicle company, a rural financial counselor, and various university or government-funded scientists. For the graziers, the forum is a chance to become familiar with changes to their industry in a format they can access as well as to raise concerns to groups that may be able to offer support. Perhaps more importantly, the forum is a chance for grazing families to meet up with each other and socialize for the first time since the wet season.

Some sessions were eagerly attended by the graziers and deemed highly useful, while others provoked heated responses. One such session was run by a group of scientists employed by the Department of Agriculture, Forestry, and Fisheries (DAFF) on a project called "Paddock to Reef." The project was intended to improve grazing management practices to minimize the amount of sediment running into waterways, out to sea, and onto the Great Barrier Reef. This was one of the final sessions of a long day, and the presenters seemed exhausted. They spoke about the benefits of improved management practices and handed around a matrix describing how graziers may be able to alter their land and cattle-management practices. The session was combative from the outset, the graziers approaching the subject with a surly kind of outrage, and the presenters seemingly already defensive, probably having encountered the same arguments multiple times that day. The presenters spoke about the "success story" of their project: a block of land, a former cattle station, that was now an entirely de-stocked conservation station. Before the scientists had finished speaking there was an outburst of protestation from the graziers gathered, with shouts of, "yeah, but who's paying for that?" and, "it all looks good on paper, but we can't very well be doing that! We need to make a living from our land." Pam was one of the more vocal graziers at this session. She pointed out that de-stocking can create other issues that do not have an impact on erosion but are problematic in other ways—predominantly, that weeds can proliferate where cattle have been removed.

The presenters tried to wrangle the group into some kind of order, seeking to bring them back to what they perceived to be the issue at hand. One presenter began talking about mapping ground cover, hoping that these maps would become useful tools for graziers. Pam responded, arguing that she had "that many maps and they're all useless." She said that researchers tended to come only at this time of year—the very early dry season, when everything was lush and green—and thus had little understanding of how the ground cover on her station changes throughout the year, whereas she frequently travels all around her lease and has done so for fifty years. She knows how her country responds to different conditions. "Well then," the presenter sighed, taking a bite of an apple. "I guess I've got nothing to offer you, Pam."

The other presenters continued trying to justify why the work and the maps may be useful, but the graziers were bristling and closed-off. A grazier named Jenny, echoing Pam's point, tried diplomatically to explain to the presenters why their approach was encountering so much resistance. She said, "this is the difference between an owner and a manager [of a station]. We are the ones who know the intimacies of the soil."

Groups like Queensland Parks and other conservation advocates, such as the presenters described here, frequently point to scientific data on the negative impacts of grazing on levels of sediment entering the waterways of Cape York in order to make an argument that grazing is not an appropriate industry for Cape York. The grazing impact with which they are particularly concerned is erosion, which is produced by soil compaction, the spreading of weeds, and reducing of ground cover. This increased erosion causes greater amounts of sediment to enter waterways that then flow into Princess Charlotte Bay and onto the Great Barrier Reef, one of Australia's most famous tourist destinations and a World Heritage site. One senior ranger for Queensland Parks told me plainly on multiple occasions that cattle are terrible for the coral reefs, that they are killing the reef, that everyone who "believes in science" knows this and that graziers willfully ignore this information because it suits them to do so.

However, the graziers and some Aboriginal people I worked with dispute these claims. Several people displayed what Gill (2005; 2014) has referred to as a belief in the gardening ability of cattle. One old Aboriginal stockman told me a story about a man he knew whose cattle had allegedly helped to fix erosion, as their dung had encouraged the growth of ground cover and gradually filled in breakaways. Another grazier told me that she believes cattle encourage the growth of plants because their cloven hooves turn over the soil, aerating it. These claims are held up alongside a general downplaying of the damage that stock inflict on the country. Still, graziers are not ignorant of the environmental impacts of cattle. As most of these graziers have been working continuously on the same station for four or five decades, they have, for instance, seen how the introduction of fences has changed the country.

Once Equal Wages legislation for Aboriginal rural workers came into effect in the 1960s, stations increasingly fenced their land—as fencing and smaller paddocks were the alternative to a large workforce that graziers could no longer afford. Around the same era, nutritional supplements came to be seen as a vital cattle management tool. As a result, cattle were able to graze on a wider variety of grasses in a smaller geographical area, leading to instances of the country being eaten out entirely, or as one grazier put it, "flogged" and "gone to shit." While graziers still rely heavily on fences and lick to have viable enterprises, experience has taught people not to overstock or push the Country beyond its capacity. One grazier, Bill, told me how quickly the Country could go from being productive and sustainable to completely depleted:

> You can see, you just watch your grasses. That joint spear and sorghum, black spear, if you see them disappear. Like, after the wet if those seeds aren't there in the paddock, you know there's too many [cattle] there. . . . Yeah, you can get rid of them native grasses pretty bloody quick. . . . Oh it don't take long to do it! If someone overstocks . . . When old mate left there [a neighboring station], sold to National Parks and the next lot come in, it was probably six years and that place still hadn't recovered properly because it had been flogged out too hard.

Having learned these lessons, though, graziers tend to think that while lick and fences can be detrimental to the Country, if managed properly the Country can continue to be cared for alongside grazing.

There are instances of landscapes that foster and allow multiple uses, alongside work to conserve and protect biodiversity and environmental values, around the world. In parts of Europe, for instance, many protected areas and parks function as what conservation biologists have called "working landscapes" (Hamilton 2018; Kirner 2015). Working landscapes allow people to dwell in and make a living off the land through various forms of exploitation—fishing, grazing, forestry, and agriculture—while working to ensure well-being for ecosystems and the variety of nonhumans that live there (Hamilton 2018; Huntsinger and Sayre 2007). The logic of protecting working landscapes is entwined with an acknowledgment that the landscapes being conserved are not "pristine nature," but have been cultivated through thousands of years of human activity and exploitation, alongside an awareness that land is cared for better when it is inhabited (Hamilton 2018; Watt 2017; Sayre 2007). Such an approach is increasingly gaining traction across Europe and in the American West. In the American West, the sizable rangelands that still exist across private ranching enterprises and tribal lands have high biodiversity and conservation values. The interest in developing working landscapes emerges from a kind of pragmatism within conservation biology. As Charnely, Sheridan, and Nabhan write, "the only way to achieve conservation at a large landscape scale is to maintain or restore ecosystem health across the jurisdictional boundaries . . . ranchers and foresters and the rural communities they help sustain have to able to make a living from the lands on which they have depended for so long" (2014, xiv). The shift toward doing conservation through working landscapes is also entwined with an acknowledgment that Indigenous people, in the Americas, Australia, and elsewhere, have exploited, cultivated, shaped, and transformed landscapes in multiple ways for millennia, producing the humanized, cultural landscapes that colonizers mistook for wilderness (Huntsinger and Sayre 2007, 4; Langton 1998, 2002; Nabhan, Knight, and Charnely 2014; Diekmann, Panich, and Striplen 2007).

Indeed, many graziers take part in a range of conservation initiatives on their stations. Cape York Natural Resource Management (CYNRM) is a local nongovernment land management organization that helps to connect graziers to various forms of government funding so that they can carry out different projects on their leases. As a result of the work by CYNRM, many cattle stations have nature refuges on their land in which particular sensitive areas, like wetlands and waterways, are fenced to exclude cattle or protect a particular threatened species. An employee of CYNRM, and the local Landcare officer, Michael possessed the kind of experience that graziers value so highly. Michael came from a career in Parks, working first in the Northern Territory and then in Cape York, but had grown up in a rural farming family. He knew how to speak with graziers and how to emphasize the common goals that CYNRM and graziers share. Importantly, Michael understood what issues graziers considered to be pressing and what kind of language to use so as not to alienate them, across issues as diverse as stocking rates, preparing for extreme weather, pest control, and fire management.

The notion of working landscapes upsets the logic that has underlain preservationist or "fortress" conservation (Brockington 2002) in America, Australia, and other states that follow the Yellowstone model for National Parks without people (Nabhan, Knight, and Charnely 2014; Hamilton 2018). By acknowledging that landscapes, ecosystems, and individual species can be cared for *alongside and through* simultaneous care for the economic and affective relationships that people have to land, a different way of envisioning care for land *other than* through conservation becomes apparent. Sheridan and colleagues write that "the term 'working landscape' implies an embodied sense of place, one that often reflects knowledge of local ecological processes that can only be accumulated through generations of daily experience with climate, water, plants, animals, and soil" (Sheridan, Sayre, and Seibert 2014, 71). This embodied sense of place, affective as well as involving rich and carefully accumulated environmental knowledge, is what grazier Jenny was referring to when she said that "we are the ones who know the intimacies of the soil."

Graziers are not perfect stewards of the land. As Bill pointed out, some people push the land beyond what it can handle, resulting in changes to the makeup of ecosystems that are sometimes irrevocable. Yet, an ethic of care runs through the work that many graziers do in their cattle and land management. This form of care is, necessarily, self-interested. Graziers want and need their land to continue to support cattle grazing so that they can continue to have a livelihood, can continue to dwell in the region, and can continue their valued way of life. But there are forms of land care enacted by graziers that go beyond utility for cattle grazing and beyond their own economic well-being. While economic relationships to cattle and land are the foundation for graziers' way of life, and ways of relating to land, tendrils of care seep out in sometimes unexpected ways. This is apparent in the willingness of many grazing families to allow scientists to visit and stay at their stations—in some instances for months, and even years, at a time—to conduct research that is related to cattle only in the sense that cattle grazing may have negative impacts on the populations and distribution of particular species. It is evident in the meticulous records that multiple graziers keep of rainfall and weather events. It is clear in the intimate knowledge of the land that graziers share, offhand and casually, during drives around their stations; in the moments during mustering when a grazier will pause to draw my attention to a particular view that they find beautiful, or at which place they have fond memories.

I suggest that what graziers are seeking to cultivate when they care for their cattle and land are "livable natures" (Scaramelli 2021, 130), working landscapes, or—as I have come to think of them—workable landscapes. By this, I mean that graziers labor to shape landscapes in which life is possible—for humans, cattle, and other nonhumans. In the notion of making something "workable" though, there is also a sense of "making do." This is a distinctly rural approach to land management and care, rooted in pragmatism, imperfect, messy, and incomplete. It is a form of care for land that works actively against the politics of purity, speaking to what has been inherited and what damage has already been wrought by poor management practices. In thinking through how graziers make do, I am reminded of Alexis Shotwell's call to aim "for a more open field of the patchy differences where we might find hope" (2016, 9). Graziers' imperfect attempts to make do, in between care and complicity, are at odds with the aspirations and commitments of Queensland Parks, which is bound by legislative mandates to enact care for land in only specific ways that do not always allow for the sprawling multiplicity and permeable boundaries that characterize life and land in Cape York.

————

Some months after Pam and Mike had discussed their anger over the delayed mustering permit with me, their relationship with Queensland Parks seemed to hit a new low point. On a visit to their station, I asked over dinner whether they had ended up mustering at Rinyirru National Park. Mike told me that they had received their permit to muster and went into the park to look for their cattle, but found no cattle left in the area in which they were allowed to muster. Instead, they came across a group of cattle who had been shot. Mike and Pam had taken photographs of the dead cattle they found. Pam said that seeing the cattle corpses "made me feel sick in my guts." Distressed enough by what they perceived to be a waste of good cattle, Mike and Pam became more upset when they realized that ears of some of the cattle were missing. They told me that it was possible that this was the work of dingoes, but they were suspicious that the missing ears were the result of someone intentionally cutting off an ear tag that would have been used to identify the stock as someone's property. "It looked like a very clean cut," Pam said darkly.

When they approached the rangers on the park about the dead cattle, Mike and Pam said that the rangers denied that any cattle had been shot in that area. Later, though, Mike spoke on the phone to the senior ranger for the whole region—the most highly ranked Parks employee that the graziers have regular access to. Mike alleges that, when confronted about the dead cattle, this individual said to him, "we told you we were going to shoot the cattle and that's what we did."

A short time later I was eating lunch with Mike and Pam's grandchild, an eight-year-old aspiring grazier named Robert. As always, Robert was dressed in a well-worn pair of jeans and oversized cowboy hat and boots, the perfect miniature of his uncles and grandfather. Robert asked where I was headed to next, as I was preparing to leave Hillview Station and move on to my next site. I told him that I would be going to Rinyirru National Park. A frown crossed his small face. "I hate Parks," Robert told me angrily. "They don't let us muster our cattle and then they shoot our cattle and cut off their ears."

The tension between Queensland Parks and graziers shows no signs of abating. Mutual suspicion shapes interactions between the two parties, and the contrasting ways in which cattle are understood remains a sticking point. Inside the boundary fence of the park, cattle are categorized (legislatively and in the management plan) as unwanted pests, causing wanton destruction to sensitive ecosystems, and requiring exclosure, control, and culling. Inside the park, the value that Aboriginal Traditional Owners place on cattle is secondary to conservation concerns. Rinyirru Aboriginal Corporation has agreed to the management plan, although disagreements remain about how best to deal with the cattle that remain on the park. Inside the park, the distinction between cleanskin and branded stock to some extent dissolves; all beasts are considered equally destructive.

On the other side of the boundary fence, the categories that different cattle are ascribed are more ambiguous—the difference between cleanskins and branded stock is one of degrees, not of discrete categories (this one has an ear tag, this one does not). Charges of environmental damage by cattle are more ambiguous too, as graziers seek to minimize the impacts of cattle on the land and emphasize their own knowledge and ability to enact care for the environment. Through a consideration of how cattle variously belong and are excluded, are valued and destroyed, the logics that underlie the different forms of care for land enacted by Queensland Parks and graziers emerges. Queensland Parks rangers are committed to a version of conservation that seeks to protect and preserve, and cultivate a landscape as devoid as possible of introduced and invasive species, and of unruly graziers. For graziers, though, embodied and pragmatic forms of care for land that can be understood as seeking to cultivate workable landscapes emerge as something other than conservation. This form of land care is messy, entangled, imperfect, and self-interested. Yet it provides a route for thinking through what environmental management might, could, and sometimes already does, look like when we reject purity as the ultimate goal (Shotwell 2016).

Queensland Parks and graziers have contrasting visions for the future of Cape York, and for how management decisions should be made today in order to bring these imagined futures into being. Different conceptions of what constitutes care for land, and how that care should be enacted, is the tension that runs through conflict over the cattle management strategy, and narratives about plans to "lock up" the Cape York region. Interpersonal issues between individuals, and the sense among both Parks staff and graziers that the other party is intentionally hostile and antagonistic, compound these tensions. There is no space within the current iteration of the National Parks model to embrace the concept of "working landscapes" to function in protected areas. At its heart, though, the struggle is one for what can be considered to be an appropriate use of land in Cape York, and whose conception of "nature," livelihood, way of life, and connection to land and place gets to matter.

PART II

Seeping

Weedy growth, Rinyirru National Park, 2019.

THREE

Weeds

On a hot, sunny, and steamy day in April 2019, I accompanied cattle grazier Mike on an expedition to spray weeds. As Mike explained to me, the end of the wet season is the optimal time to spray weeds as many are only just popping up in the wake of the monsoon and they have not yet gone to seed. If weed control happens after the plant has seeded, it is already too late. Mike's focus was on a weed called sicklepod. Sicklepod (*Senna obtusifolia*) is classified as a woody weed. It is a variety of the Senna group of species—plants from America that have been spreading voraciously through Cape York. Sicklepod shrubs grow up to two meters in height and have soft, rounded green leaves and attractive yellow flowers. It is a competitive weed that quickly takes over large swathes of land, spread by livestock, tire treads, boots, and water. Sicklepod seeds remain viable for about ten years, making the plant very difficult to control. Graziers generally target woody weeds like sicklepod for control rather than introduced grasses because sicklepod reduces pasture, whereas cattle can eat some introduced grasses.

Mike and I took out two quad bikes and traveled to some parts of the station where sicklepod was becoming problematic. While Mike rode his bike in a methodical kind of grid formation, spraying weeds from a tank of poison on the back of his bike, I waited in the shade. Mike told me that he just wanted me there to help him open a few big and unwieldy gates and in case he went "arse over head" on his bike. Mike told me that, in the past, he used to attack sicklepod more aggressively, spraying in any area on his lease where he was aware it had cropped up. However, the Cook Shire Council, which assists many landholders in weed control, had stopped focusing on spraying sicklepod, and Mike was not able to keep up a concerted campaign alone. Instead, he focused on spraying weeds just around the yards and lick[1] sheds. These are places where he needs easy access and so prefers to keep "clean" of the weed. Mike said that he knows he will never beat sicklepod; all he can really do is clean up the areas he needs to have clear. Pausing from spraying to refill the tank on his bike with water from a nearby creek, Mike told me that he was feeling particularly pessimistic about his chances of controlling the weed, as a series of medical appointments had kept him down south in Cairns for some weeks, leading him to miss the optimal time to spray. He explained that the sicklepod may have already gone to seed and he was concerned he was fighting something of a losing battle.

On a nearby station, grazier Alan expressed a similar cynicism about his ability to control weeds. He told me that he had given up on even trying to control sicklepod in recent years. After a morning of repairing fence lines and floodgates with Alan and his wife, Bev, after the monsoon, we drove into a paddock that had been utterly transformed from the dry. Where a few months before there had been an empty and dusty paddock with limited ground cover and a couple of soapbush trees, there was now a sea of green. Sicklepod, growing to above head height, had overtaken the paddock. In Alan's car, we edged slowly through the plants so that he could check how much water was in a dam in the center of the paddock. He joked, saying he hoped he could remember a route with no logs in our path. Alan explained to me that until recently he would use between 18,000 and 20,000 liters (4,755 to 5,283 gallons) of poison each year to control sicklepod. When I asked why he had stopped spraying weeds, he told me that the Lama Lama rangers who do land management on his station "don't come to the party." During the wet season, the Lama Lama rangers are on a break from work and so they only start spraying weeds on Alan's lease around July or August, when the large team of casual rangers come back to work. By then, the plants have already seeded and there is little point in spraying.

"You may as well just slash the weeds," Alan said, adding that he thought spraying at that time of year was nothing but "a waste of good poison."

Living among and against Weeds

The spread of invasive plant species is of increasing concern for all land managers in Cape York. This chapter explores how people labor in and against a landscape beset by weeds. In Cape York, weeds seep into places, and across the porous boundaries between land tenures. The problem of weeds is overwhelming, pervasive, and seemingly intractable. People kill weeds to enact multiple forms of care, for landscapes, ecosystems, native species, livelihoods, visual amenity, ancestors. But spraying weeds is not a straightforward practice with predictable results. The chemicals used to spray weeds have their own hard-to-quantify impacts, and the timescales that weeds operate in don't fit easily with funding cycles, with the result that land managers may not be adequately resourced to focus on a species for the length of time that it would take to achieve results. Killing to care involves a trade-off. And, despite enormous effort, land managers enact this care in the shadow of very little success.

Like much of northern Australia, Cape York is particularly threatened by the spread of gamba grass (*Andropogon gayanus*) and grader grass (*Themeda quadrivalvis*), species that were initially purposefully introduced in order to provide pasture for cattle and whose ecological impacts—particularly on the viability of current burning regimes—have become increasingly evident in recent decades (Neale 2019). It is the control of these weeds that groups, like Cape York Natural Resource Management[2] (CYNRM) and the Cook Shire Council,[3] are directing funding toward.

Weed control is a significant task for cattle graziers, Queensland Parks, and Aboriginal ranger groups alike. As introduced plant species are not always constrained by fences or land-tenure boundaries, land managers are aware that controlling these species must be a cohesive and coordinated effort. Different land managers target particular introduced species for control in the areas for which they have responsibility, and the ways that certain species are understood as more or less problematic indicates the ways in which land managers understand and seek to order landscapes. For all land managers, weed control embodies a form of caring for the landscape that is interested, engaged, and involved. It is bodily, and noninnocent. Weed control entails a form of violence, a variation of killing as a form of care.

There is a tendency in recent scholarship to pull apart and dismantle the categories by which we order things (as native or invasive, for instance) and to see "weediness" or hybridity in the landscape as something generative, a thought experiment productive for thinking through how to combat the overwhelming emotional weight of the Anthropocene (Bubandt and Tsing 2018; Doiron 2023; Tsing 2019). However, as Jessica Cattelino has suggested, the task for social theorists is to "not just break down or destabilize categories but rather analyze what sustains them, and with what political and economic effects" (2017, 131). In thinking through invasive species management, I am caught between a desire to move away from and critique the notion that native species are coded as good, and exotic species coded as bad, and an unproblematized embrace of weediness as generative. In a sense, it is possible to think around or away from this bind by reshaping the categories that we use to determine management practices. Instead of native versus exotic, Lesley Head (2012) suggests it is more productive to think about the behavior and status of a species. She suggests that a better metric is invasiveness (as a behavior) and vulnerability (as a quantifiable status). In many ways this is what is already happening in conservation and land management, as practitioners with limited time and resources must determine what projects are more and less urgent. As will become clear throughout this and the next chapter, the values that underpin what different people deem to be an urgent project differ depending on a complex set of values and concerns linked to livelihoods and lifeways.

Plants have long figured significantly in the intersection of social and ecological research, particularly in the fields of ethnobotany and economic botany, but also political ecology (Robbins 2007). As Cori Hayden points out, "the study of plants and knowledge about their use has a long and complicated legacy in which resource extraction has unquestionably played a prominent role" (2003, 30). In much of this scholarship, plants enter the social world as resources that have the potential to draw capital and transform the lives, livelihoods, and homes of local peoples. However, proponents of the recent interdisciplinary "plant turn" argue that social and philosophical theorists have failed to seriously consider what Michael Marder calls "the logic of vegetal life" or what Jenny Atchison and Lesley Head have referred to as the "plantiness" of plants (Atchison and Head 2013). Thinking with plants, by attending to their "plantiness" means considering a set of "distinctive capacities" (Head, Atchison, and Phillips 2015). According to Head and colleagues, part of what makes these capacities distinctive is that plants are what they call "relational achievements"; they come into being through relationships between elements, the soil, and a whole host of other beings. For thinking about weeds, though, the prescient elements of plantiness are plants' capacity for movement independent of human meddling and what we can think of as plant time (Head, Atchison, and Phillips 2015). Considering how weeds shape social worlds means paying attention to how plants move in time at multiple scales simultaneously; what Marder (2013, 107) calls heterotemporalities. Moving at different timescales to humans, invasive plants confound land managers' attempts at control, particularly for those land managers whose cycles of work are driven not only by seasons and need, but by bureaucratic structures and systems, by institutional mandates, by key performance indicators. It is in the clash between environmental governance and weeds that the stakes of these heterotemporalities are laid bare.

Weed Control in the Park

Weed control is also a significant activity for the Aboriginal and non-Aboriginal rangers in Rinyirru National Park. When I spent time in the park during the wet season, the focus was firmly on lion's tail (*Leonotis leonurus*). Lion's tail is a woody weed from southern Africa. The shrub can grow up to two meters tall and has dark green leaves with serrated edges and stunning orange flowers. Lion's tail is sold as an ornamental shrub in many parts of Australia. It thrives in disturbed areas and along waterways. Being able to spray lion's tail depends on the weather, as any rain in the hours after the poison has been administered will wash the plants clean and render the poison useless.

One wet season afternoon, I clambered into a buggy with Ray, the ranger in charge for Rinyirru National Park. He passed me a protective face mask and instructed me to put it on so as not to ingest any of the chemical that we would be using. We were on our way to spend the afternoon spraying lion's tail. Ray sped down the main road through the park before cutting into a section of bushland. He had a GPS in hand as we zigzagged through the scrub to a plotted coordinate where lion's tail had previously been spotted. We reached a clearing dotted with tall shrubs. Ray pointed out the plant to me, describing some of its notable features. He passed me the nozzle of a hose connected to the tank of poison on the back of our buggy and showed me how to spray the weed depending on proximity and the wind. A gentle mist that will coat those plants that are nearby, and a steady jet for those plants farther away. Even through my mask, the poison had a bitter smell. It fused with the scent of wet earth from the frequent rain, crushed grass from our bushbashing, and the exhaust fumes of the buggy. Once the section had been sprayed, we marked the area on a GPS and tied a pink plastic ribbon around a nearby established tree to ensure that the other teams of rangers would not accidentally double up and spray this section again.

It took a while to get the hang of spotting the weed. Sitting beside Ray in the buggy, I rode with him slowly through clearings filled with vegetation, looking for the telltale leaves. At first, I frequently mistook other plants for lion's tail, not able to differentiate variation in leaf texture from a distance. After some time, though, I was able to correctly identify the plant, spotting it among a sea of green.

We spent weeks traversing open clearings and thick and swampy undergrowth, encountering multiple punctured tires and becoming bogged in sticky mud several times, as we sought out and sprayed lion's tail. Usually, once one plant was spotted it became clear that a whole area was inundated with the plant. It was hot, sticky, and uncomfortable work. In the wet season, it's not uncommon for the temperatures in Rinyirru National Park to be in the high thirties (Celsius—equivalent to around 100 degrees Fahrenheit) and for the humidity to sit at around 98 percent. The face masks—necessary to avoid breathing in too much herbicide—contributed to a stuffy, almost claustrophobic sense, heightened by the dense vegetation we were moving through. One of the younger rangers, Ivan, sometimes brought a small Bluetooth speaker with him and listened to music to pass the time. Other rangers chatted casually, recalling specific incidents in the park or describing particularly good fishing spots. As with most wet season tasks, the rangers seem to find some relaxation and joy on these weed spraying expeditions—despite the difficult physical conditions.

But with weed control, it's difficult to appreciate whether such laboring yields sufficient results. On an afternoon spraying the weed with Ray, I asked why lion's tail was the focus. Ray replied that he didn't really know why they spray lion's tail in particular, given that Queensland Parks has been targeting the plant for years and have had very little success. He said that they are trying to contain it, or at least slow the spread. On that day, as Ray and I negotiated and sprayed seas of lion's tail, he said despondently that the patches we had come across were the worst he had ever seen, much larger than in previous years.

The rangers' ongoing, long-term attempts to control lion's tail highlight the conflict between a vegetal timescale, what Margulies (2023) calls "phytotemporality," and bureaucratic land management and environmental governance. Moving too slowly to be apprehended by humans, and too fast to be easily contained and controlled, invasive plant species confound human attempts to manage them. When a lion's tail plant goes to seed, hundreds of seeds are spread through waterways, in mud that adheres to animals and clothing, and in tire treads. Seeds can be spread in an instant, and can then remain dormant in the soil for years, waiting for the correct conditions to grow. According to the rangers, each lion's tail seed pod contains 400 seeds that have a germination success rate of between 98 and 100 percent, and the seed bank lasts for seven years. These seed banks, an almost spectral presence in the soil, demonstrate how rangers are working in and with an inherited landscape, shaped by distant and more recent pasts.

Over an afternoon beer back at the ranger base (a time-honored wet season work tradition), Ray noted that eradication is an unrealistic goal, given that this would require spraying every single lion's tail plant for seven years without a single plant going to seed. Ivan, who had previously worked for Biosecurity Queensland[4] joked that "eradication is a dirty word." He said that he was pessimistic about the possibility of eradicating any weed. Some of the older rangers present told me that they had been trying to control lion's tail for twenty years now, and the "footprint" of the weed (the geographical area where it has been found) continues to grow larger. We sprayed the lion's tail consistently for weeks, and these outings frequently provoked questioning from the rangers about how effective we were. Ivan told me that he had spoken to a botanist employed by Queensland Parks who believed that the rangers should not bother spraying lion's tail at all given that they were not achieving any progress with it. The ranger who had been at the park the longest, Roger, chipped in and said that probably all we were achieving was to slowly give ourselves cancer from the poison.

Graziers and Queensland Parks alike invest significant amounts of time, labor, and money into attempting to control weeds. While the purposes of their respective weed control programs diverge, graziers and Queensland Parks categorize plants in similar ways as either useful, appropriate, and belonging in some sense, or as problematic, invasive, and requiring action.

Defining "Weeds"

Invasive species management is considered an urgent issue across much of Australia (Kull and Rangan 2008; Atchison and Head 2013; Gibbs, Atchison, and Macfarlane 2015). In such a context, weed control is often framed through the lens of war or a battle, understood as an "epic [drama] of ecological assault by alien plants over native species" (Kull and Rangan 2008, 1259; Atchison and Head 2013; Bach and Larson 2017; Radomski and Perleberg 2019). Plants have always traveled across landscapes and permeated borders, sometimes purposefully dispersed by humans and sometimes seeping into new geographies without human intervention. How these introduced species are understood, categorized, and valued by people is culturally and ideologically contingent, rather than reflecting any kind of biological fact (Robbins 2004; Kull and Rangan 2008; Head 2012; Doody et al. 2014). Definitions of weeds are slippery, with weeds loosely described as plants that are out of place, but often being roughly mapped onto a native/nonnative binary (Qvistrӧm 2007; Doody et al. 2014).

Geographer Qvistrӧm has argued that "weed is not a botanical concept; rather it is defined as any plant in the wrong place, and therefore a problem or at the very least a plant of no value" (2007, 272). Bubandt and Tsing define weeds simply as "organisms the proliferate without human planning" (2018, 2). They suggest that the term "weed" need not be a negative ascription, saying that "the weeds we identify may be good or bad to the humans amidst whom they thrive. Although they are unplanned, they may become resources for humans; alternatively, they may hamper resource utilization—or both" (Bubandt and Tsing 2018, 2). The concept of "weed" works as a kind of metaphor, then, used to describe a landscape that is disordered. Of course, while the definition of weeds may be slippery and contingent, plants designated as weeds in Cape York do have tangible effects on ecosystems and economic futures, with the spread of weeds impacting biodiversity, visual amenity, the availability of pasture, and the viability of fire regimes. Importantly, weeds (and their impacts) are not static. Bubandt and Tsing (2018) note that, although many plants now considered "weeds" were initially introduced deliberately by humans, the movement and mobility of such species means that they come to exist in ways that are unexpected and, frequently, problematic for humans. As they write, "humans are given back landscapes differently than the ones they imagined and sought to make" (Bubandt and Tsing 2018, 2). Such unintended collaborations that emerge as a result of these interactions between humans and plants ("feral dynamics") precipitate human attempts to control, discipline, and reorder landscapes.

For land managers in Cape York, I suggest that weed control is ultimately an attempt to reorder landscapes in a way that encourages certain landscape attributes that are preferable to particular groups for specific reasons. Graziers seek to control weeds in order to maintain pasture for cattle grazing, an aspiration that is tied to both economics (as cattle are dependent on pasture) and to carrying on a valued way of life (as the social category of "grazier" is reliant on grazing). For their part, park rangers are following policy guidelines that require them to work toward cultivating a "natural"—which we can read as "native"—landscape. Aboriginal rangers tend to position weed control as part of a broader project of caring for the landscape and the ancestral spirits that inhabit it. Other stakeholders, such as CYNRM, Cook Shire Council, and the federal and state governments, are concerned with ensuring the viability of fire regimes in order to maintain the carbon sequestration program and thus safeguard the economic future of Cape York.

To disentangle these preferences and specific reasons we not only need to deconstruct the categorization of "weeds"; the concepts of native and nonnative also require interrogation. Even in the biological sciences there is ambiguity around the categories of native and nonnative. Head, citing Chew and Hamilton, notes that the concept of biotic nativeness is "theoretically weak and internally inconsistent" and does not stand up to empirical scrutiny (2012, 168). She argues that "when analyzed closely, characterizations such as nativeness tell us more about human bounding practices than anything inherent about the plants and their evolutionary processes" (Head 2012, 171).

Nativeness is often equated with belonging, but it is a correlation that requires interrogation (Trigger 2008; Robbins and Moore 2012; Bach and Larson 2017). The distinction between native and nonnative is based on a temporal threshold, which in Australia is 1788, the year that Britain established a colony on the continent. This temporal threshold is problematic if we consider colonization to be a structure, not an event, and furthermore recognizes neither the multiple temporal "frontiers" of colonization in Australia, nor the agency of First Nations people in Australia in terms of species distribution (Trigger, Toussaint, and Mulcock 2010, 282; Martin and Trigger 2015). Yet, there is a certain moral "good" assigned to native species and a negative agency assigned to invasive species—with some notable exceptions (Einarsson 1993; Robbins 2004; Trigger 2008; Gibbs, Atchison, and Macfarlane 2015; Brice 2014).

In Cape York, the way that plants are categorized as belonging or in need of control is related more to the behavior of the plant, and its potential utility, than to its status as native or introduced. For instance, the introduced species of mango, banana, tamarind, and lemon that grow throughout Cape York are ambiguous but relatively uncontentious. These are species valued by a diversity of land managers because of their use as a food source, but not targeted for control. Here, the arbitrary categories of native and nonnative are not deployed and instead, the behavior and possible use of the plant species in question is what is important (Head 2016; Kull and Rangan 2008). These fruit trees are introduced, but they do not tend to spread in a way that is uncontrolled and problematic. Instead, the plants that are attributed negative agency are species like lion's tail, sicklepod, and gamba grass—plants that variously threaten livelihoods, conservation values, and amenity. Managing weeds is about creating order in a landscape (Qviström 2007). Such "ordering" can be understood as preserving or shaping an "appropriate" or "natural" landscape. What counts as natural or appropriate differs from land manager to land manager, and as such it is pertinent to critically consider what the "baseline" is that we seek to return landscapes to (Trigger 2013). Such a baseline becomes particularly problematic if it is an imagined, prehuman, and pristine wilderness that land managers are seeking to produce (Cronon 1996; Robbins and Moore 2012; Shotwell 2016).

Weed Control and Preferred Landscapes

While only introduced species with invasive behaviors are targeted for control, different species are prioritized by people for specific reasons. In the case of Queensland Parks, a mixture of factors determine which weeds are targeted for control. These factors include a departmental requirement to target specific species that have been identified as urgent by Biosecurity Queensland, the observations by on-the-ground rangers and the region-wide weed management team about weed proliferation, and the outcome of joint management meetings with Aboriginal Traditional Owners. Queensland Parks have obligations to engage in weed control under the Biosecurity Act 2014, the Nature Conservation Act 1992, and the Forestry Act 1959 (Queensland Government 2017). While these priorities are geared toward an outcome in which the spread of invasive species is mitigated or stopped, the actual practice of controlling weeds remains important for the institutional culture of Queensland Parks. For Queensland Parks rangers, there appears to be importance placed on being "seen to be doing something about weeds" (Atchison and Head 2013, 961), even as rangers remain uncertain about the possibility of success with weed control programs. This is so even where such efforts are understood by rangers to be having little effect—as with the spraying of lion's tail. For Parks, "performance" tends to be measured in terms of activities carried out (the use of a particular amount of poison, the geographical area that is sprayed, the number of days the rangers spend spraying weeds) rather than about the outcome of weed control programs on the spread of weeds themselves. In a sense, this kind of organizational structure points out the impossibility of containing weeds, and the ways in which the plants elude control and containment, as well as the difficulty involved in monitoring the spread of invasive species across such a vast and remote geographical area. Perhaps more to the point, though, this is an indication of something about the culture of "the way things are done" in Parks.

While a park without weeds is not possible, the effort to control and aspiration to eradicate weeds is about the preservation of a particular type of preferred landscape. A key aim of the Nature Conservation Act 1992 is to protect "the biological diversity of native wildlife and its habitat" (s. 5d). Biological diversity is here defined as "the natural diversity of native wildlife, together with the environmental conditions necessary for their survival" (s. 10.1). While there is no specific definition given for invasive species, weeds, or ecosystem biodiversity in this act, the act is geared toward preserving a loosely defined "natural diversity" of species (s. 10.1). From the actions of Queensland Parks rangers, though, a preferred landscape would seem to be one devoid of introduced plant species. While the various pieces of legislation that mandate the control of invasive species for Queensland Parks are geared toward protecting biodiversity, in practice weed control serves multiple purposes.

For some Queensland Parks rangers, the control of weeds is related to preserving the amenity of National Parks for the visual enjoyment of tourists. Such a point was illustrated one evening when a senior ranger arrived at the base having driven around the northern section of Rinyirru National Park. Somewhat defeatedly, she spoke about the proliferation of gamba grass along the roadside near a section of the park called Nifold Plains. In a similar vein to the introduction of buffel grass (*Cenchrus ciliaris*)[5] in the Northern Territory and Central Queensland, gamba grass was initially introduced to provide pasture for cattle and has since spread widely throughout northern Australia. Gamba grass now represents a threat to the continuance of existing fire regimes due to the grass having a higher oil content than native grasses and, thus, causing hotter and more out-of-control wildfires than native grasses. The current situation with gamba grass represents the kind of "feral dynamics" discussed by Bubandt and Tsing (2018).

The spread of gamba grass is, ultimately, the result of a human and nonhuman collaboration, a nexus of cattle, graziers, the Department of Primary Industries (which encouraged planting gamba grass as pasture in Northern Australia until very recently), the Commonwealth Scientific and Industrial Research Organization (which bred the specific cultivar that is now present in Northern Australia), the landscape, and the climate, which has resulted in reduced native grasses and an increased risk of hot fires—threats that make the already contingent economic future of Cape York even more precarious (Cook and Dias 2006; Head and Atchison 2015). In their work on gamba grass spread in the Northern Territory, Neale and Macdonald have described the proliferation of gamba grass as a "slow disaster," drawing particular attention to the risk the grass poses to the continuation of the carbon sequestration scheme (2019). While these aspects of gamba grass are of great concern to Queensland Parks, on this occasion the senior ranger framed her frustrations around the impact the grass may have on the attractiveness of the park. One of the more visually stunning parts of Rinyirru National Park, Nifold Plains is an open expanse, dotted with towering termite mounds and little else. Photographs of the plains often feature in tourism brochures and four-wheel driving guides to the region. "What are tourists going to think if they come all this way to see the plains and can't see anything through the wall of gamba grass?," she asked rhetorically, frustrated. In this way, gamba grass threatens the future viability of tourism, a key, emerging industry for the region. As the senior ranger's comments indicate, Queensland Parks is concerned with ensuring that tourists visiting the park encounter the kind of landscape they are expecting. The experience presented to tourists, then, is primarily a visual one, in which the "natural" landscape—continuous with how it is presented in tourism brochures and pamphlets—of Cape York can be observed.

Whether provoked by a desire to protect biodiversity or preserve the attractive features of National Parks, weed control by Queensland Parks represents the valuation of a particular type of "natural" landscape. This is related to the preservationist model that National Parks are based on, and the notion that such places are sites of pristine "wilderness," requiring protection from the polluting impacts of humans. As noted in the introduction, this reading of landscapes as wildernesses necessitated the exclusion and erasure of people who used and relied on such landscapes for subsistence. Indigenous people were erased and excluded by the Yellowstone model of National Parks, and this practice of dispossessing Indigenous peoples for the purposes of conserving landscapes is evident across the world in a variety of contexts (Tsing 2005, 100; Doolittle

2005; West 2006). Even in contemporary situations of formal comanagement of National Parks by Queensland Parks and Aboriginal corporations, where there is legal recognition that the presumed "wilderness" of Australia is actually a landscape that has been actively managed by Aboriginal people for tens of thousands of years, these persistent ideas about wilderness tend to endure and inform some Queensland Parks practices (Adams and Mulligan 2002; Langton 2002; Slater 2013). As pointed out by Slater (2013), it is such environments' proximity to "naturalness" that designates them as worthy of care and concern.

The exclusion of Indigenous people from this model of conservation is significant because it negates the human role in the distribution of species and creation of landscapes. These exclusions helped to create the myth of a "natural landscape" untouched by human hands. However, like almost everywhere on earth, landscapes in Cape York are anthropogenic landscapes, shaped by millennia of human activities, care, exploitation, and labor (Langton 1998). More recently, landscapes in Cape York have been altered by the pastoral industry. Shotwell (2016) has argued that environments all over the world are already contaminated. She writes that "we are all living after events that have changed, and frequently harmed, ecosystems and biospheres" (Shotwell 2016, 9). Rangers today are working with and against an inherited landscape, one that has been shaped by decades of cattle grazing, altered fire regimes, and the introduction of new species.

However, not only is this landscape inherited by rangers, it is also constituted by them. In the course of their day-to-day work, rangers drive through the bush, sometimes along barely perceptible tracks, sometimes following no track at all. These diversions are necessary to carry out a range of tasks; being such a large and remote park, not all sites are accessible by well-maintained roads. Tasks like monitoring and maintaining fences require rangers to force their way through scrubby undergrowth, sometimes in smaller vehicles, like a type of all-terrain vehicle called a Polaris, but more often than not in their Landcruiser Workmates—large, boxy, utility vehicles. In their cross-country driving, rangers carry seeds in the tire treads of their cars. As they walk through the bush, seeds attach to the fabric of their clothes and socks. This is acknowledged as an issue. On a drive to check a fence line with the ranger in charge for Rinyirru National Park, he (Ray) admitted to me that what we were doing was probably contributing to the spread of invasive plant species in the park. There are attempts to mitigate this. At the end of each ten-day shift, the rangers will thoroughly wash the vehicles that they have been using. Yet there is an acceptance among the rangers that despite their best efforts, despite the care with which they carry out their work and seek to minimize any negative impacts, they will continue to play a role in the spread of weeds. Complicity and care, here, intermingle. Rangers labor toward a park free of weeds while simultaneously acknowledging how their own actions contribute to the impossibility of this goal. Spraying and spreading weeds alike, rangers nonetheless persist in this laboring. In an imperfect situation they make do, enacting these partial and complicated forms of care for the park as, simultaneously, an imagined wilderness, a series of sensitive bioregions, and as a state government project of orderliness and control.

Weed Control as Caring for Country

Another perspective indicating a different kind of preferred landscape emerged from my discussions with Rinyirru Aboriginal Corporation ranger and traditional owner Donna. As with all of the Land Trust rangers, Donna was only employed during the dry season—the result of complex funding arrangements. Upon her return to the park after the wet season, she identified several areas that she had previously worked to keep clear of weeds. Seeing the growth of these weeds, Donna told me that she thought the proliferation of weeds in the park was "disrespectful to the land." For Donna, this disrespect extends to the old people who inhabit the landscape, the ancestral spirits who dwell there. Her comments about the state of these areas now, in the wake of her absence, indicated that she perceived weed spread in these areas to relate to a lack of care and attention to changes in the landscape on the part of Queensland Parks rangers. There is a sense—here and elsewhere—that the spread of weeds can indicate to people that the landscape has "deteriorated" ' and been "abandoned" (Førde and Magnussen 2015, 190). Similarly, Bach and Larson (2017) have described how, for Indigenous elders in the Kimberley, the presence of weeds could indicate that a landscape was "sick" or "down," because local Aboriginal people had not fulfilled their responsibilities to care for Country.

Invasive grasses, near Cooktown, 2018.

Nifold Plains, Rinyirru National Park, 2019.

Donna seems to understand weed control as an alternative method of "keeping the country 'clean'" (Rose 1992, 106), a responsibility for Aboriginal Traditional Owners that is normally associated with burning the landscape. Just as burning and regularly visiting tracts of land are understood by Aboriginal Traditional Owners as significant in maintaining a connection with a landscape and the "old people" (ancestral spirits) dwelling there, so too can weed control emerge as a way to "let the country know that people are there" (Rose 1992, 106). By controlling the weeds in this area, Donna was demonstrating care for the land and her ancestors. Just as regularly visiting, camping, using, and burning the land serves to reaffirm connections between and care for the land and ancestors (Martin 1993; Von Sturmer 1978), so too does weed control function as an extension of Donna's responsibilities and ongoing relationship to the land. Donna was not alone in conceiving of weed control as caring for Country. Another Aboriginal traditional owner and ranger, Leanne, told me on a number of occasions: "rubber vine is my passion." Rubber vine (*Cryptostegia grandiflora*) is a creeping parasitic vine originating in Madagascar that has spread along waterways in Cape York, smothering native vegetation. Of course, when Leanne said that rubber vine was her passion, what she meant was that killing rubber vine was her passion.

Spraying weeds is not an innocent act. It is a form of care for native species, ecosystems, and ancestral spirits that involves some form of violence—even if the violence is being enacted on plants that we may not consider the same level of moral subject as an animal. This care is enacted through material practices: through visiting particular sites, monitoring and mapping the footprints of weeds, and controlling weeds using herbicides. While the herbicides Queensland Parks uses to control weeds are targeted to reduce impacts on other species, spraying sicklepod, lion's tail, or rubber vine does not occur in a vacuum. Of course, spraying weeds is only one way to manage them. Organizations with greater resources—such as Biosecurity Queensland—are able to facilitate their employees walking the landscape in a line, removing plants by hand and disposing of seeds in a meticulous way. This requires an enormous amount of labor, time, and money—resources that are not available to the same extent to Queensland Parks or Aboriginal land managing organizations who must juggle a number of mandates and priorities while managing large tracts of land with relatively small workforces.

Anthropological work on weeds that considers them as a metaphor, simply a plant out of place, does not necessarily leave analytical space to grapple with the very real environmental impacts that weeds can have in Cape York. And for Aboriginal Traditional Owners like Donna and Leanne, weed control is not just about reducing environmental impacts to protect biodiversity but is also about caring for the "old people" or ancestral spirits who inhabit the landscape, about keeping the land socialized, about letting the old people know that Traditional Owners are present and are enacting care.

Donna's focus on weed control as a method for fulfilling her obligations to the land and "old people" has emerged from the context of intercultural knowledge sharing she has encountered in her work alongside Queensland Parks rangers. While her desire to care for land through physical engagement predates her employment at Rinyirru National Park, it is through her work as a ranger and her training in a Western scientific form of land management that weed control has emerged as one of the ways in which she understands caring for Country to occur. For Donna, then, controlling weeds is less about preserving a particular kind of landscape than it is about maintaining a particular kind of relationship to the land. In engaging in weed control, Donna—like the Queensland Parks rangers—is concerned with the actual practice of weed control, with being "seen to be doing something" (Atchison and Head 2013, 961). However, in "doing something" she is seeking to show care and respect to the "old people" of the landscape, whereas Queensland Parks staff are attending to departmental, funding, and management requirements. For those Queensland Parks staff who are also Aboriginal Traditional Owners for the park, these differing priorities blur as, through weed control, these rangers fulfill their obligations to the landscape, their ancestors, and the Queensland government simultaneously.

Weed Control and Grazing

The way that graziers approach weed control indicates a similar aspiration to preserve a particular kind of landscape as exists among Queensland Parks rangers—although in the case of graziers, this preferred landscape is a pastoral one. For graziers, the actual act of weed control is less significant than the outcome. Graziers seek to ensure that they have sustainable pasture in order to continue running cattle. Hence, the focus on woody weeds over invasive grasses is logical for graziers as it is woody weeds, not introduced grasses, that threaten their pasture. While acknowledging that these introduced grasses do pose a threat, especially in terms of affecting burning practices and bushfire risk, graziers assert that because cattle eat these grasses, they do not require immediate attention, and they may even provide some benefit. Some graziers suggested that introduced grasses, like gamba grass, actually mitigate erosion, particularly along waterways. In the sense that it can function as pasture and to mitigate erosion, gamba grass can be considered by graziers to be a "useful plant," rather than a "problematic weed" (Kull and Rangan 2008, 1270). Where invasive species are understood to hold utility, coexistence with such species becomes possible (Atchison and Head 2013). Graziers are thus more concerned with the spread of woody weeds that compete with and ultimately greatly reduce pasture.

The ways in which graziers value land and understand their place in the landscape of Cape York is evident in their efforts to preserve and protect pasture. As with Donna, graziers are concerned with maintaining a particular relationship to land; one that is based on a valuation of their economic relationship to land and embodied sense of place. Graziers' relationships to land are based on their ability to work in, on, and with landscapes. The economic viability of their industry in the region is reliant on their ability to produce pasture that can support cattle. Their weed control is aimed at ensuring this. But with weeds like sicklepod not drawing the attention of other land management groups in the region, graziers are limited in their ability to mitigate the spread of such weeds. In the same way that gamba grass represents a "slow disaster" for the viability of the carbon sequestration scheme (Neale and Macdonald 2019), so too does the spread of woody weeds represent a gradual threat toward the viability of the grazing industry.

Just as animals like cattle and sheep have functioned as colonial agents, so too have introduced plants been used to reconstruct landscapes for political and economic purposes, sometimes deliberately and sometimes unexpectedly (Dominy 2003; Galvin 2018). Plants in Cape York also "do things" (Galvin 2018). Gamba grass, introduced deliberately and initially planted as pasture for cattle at the encouragement of the Department of Primary Industries (Atchison and Head 2013; Petty 2013), disrupts existing fire regimes with its high fuel load. Sicklepod, introduced as a green manure crop, competes with pasture, threatening graziers' livelihoods and exacerbating tensions between graziers and Aboriginal ranger groups, Cook Shire Council, and CYNRM. Lion's tail, an escaped introduced ornamental shrub, competes with native grasses and threatens biodiversity. Different weeds and their various impacts make precarious the established forms of land management that exist in the region, particularly for graziers. In this sense, plants "bring into being" (Galvin 2018, 243) new frictions, tensions, and relationships among humans and between humans and landscapes. As the spread of invasive plant species continues, aided and abetted by a changing climate and an increasingly mobile human community as tourism to the Cape becomes increasingly accessible, weeds and land managers are likely to become more entwined.

———

The categories by which nonhuman species are ordered are culturally and socially important, and, as Head (2017) has suggested, can reveal much about human bounding practices. Through the deployment of categories like native and nonnative, introduced and invasive, feral and pest, species are situated as belonging or not belonging in a region. The ascription of these categories spreads into practices that work to order landscapes in particular ways, related to how different groups of people understand, rely on, and relate to land.

As with all landscapes on earth, Cape York is a "hybrid landscape" (Mulcock and Trigger 2008) impacted by climate change, human activity, and the global movement of resources and species. In addition, as Bubandt and Tsing have suggested, some anthropogenic landscapes are not solely of human design but emerge as a result of "the cascading effects of more-than-human negotiations" (2018, 1). Lion's tail was intended as an ornamental garden shrub, sicklepod as a green manure crop, and gamba grass was originally intended to provide stock feed. Each of these species has seeped beyond the confines of its imagined and intended existence, trickling into new geographies and ecosystems and accumulating "weediness" along the way. Bubandt and Tsing refer to these spaces of new kinds of unexpected relating as imbued with "feral dynamics" (2018).

Cape York is a place rife with these feral dynamics, as humans and nonhuman species are brought into unintended relationships that sometimes represent collaboration and often result in human effort to exert control. Such relationships are frequently informed by economics; as in Cape York landscapes, labor and livelihoods are intimately entwined. When people go to grapple with invasive species, their economic relationship to land is part of the grounds of encounter. This is because it is through making a viable living on and with the land that people are able to form meaningful attachments to land and craft meaningful senses of belonging. The categories that different species are ascribed differ from land manager to land manager, just as notions of what a preferred landscape or appropriate use of land is changes depending on who you ask. Relationships between humans, nonhuman plant species, and landscapes are contingent, contextual, and continually coproduced in intercultural and interspecies assemblages.

In enacting care for landscapes, some species are deemed threatened while others are deemed threatening. In weed control in Cape York, particular invasive species are targeted for control. In killing these weeds, land managers are engaging in acts of care toward sensitive ecosystems and landscapes. Killing to care, here, means attending to the "heterotemporalities" (Marder 2013) of invasive plants, considering and responding to the multiple timescales at which they operate, and persisting in continuing to enact caring labor, through the application of herbicides, the GPS marking of the spread of plants, and the slashing of weeds, even when the impacts of this caring labor are difficult to apprehend. For Aboriginal Traditional Owners, this care extends to the ancestral spirits who dwell in the landscape. By controlling weeds, they are keeping the Country "clean" and demonstrating ongoing care for their homelands. For graziers, controlling weeds is entwined with care for cattle, a form of care that is implicated and entwined with their livelihoods and ability to remain living in the region. Weed control, then, is a practice that draws in and enfolds multiple kinds of caring labor.

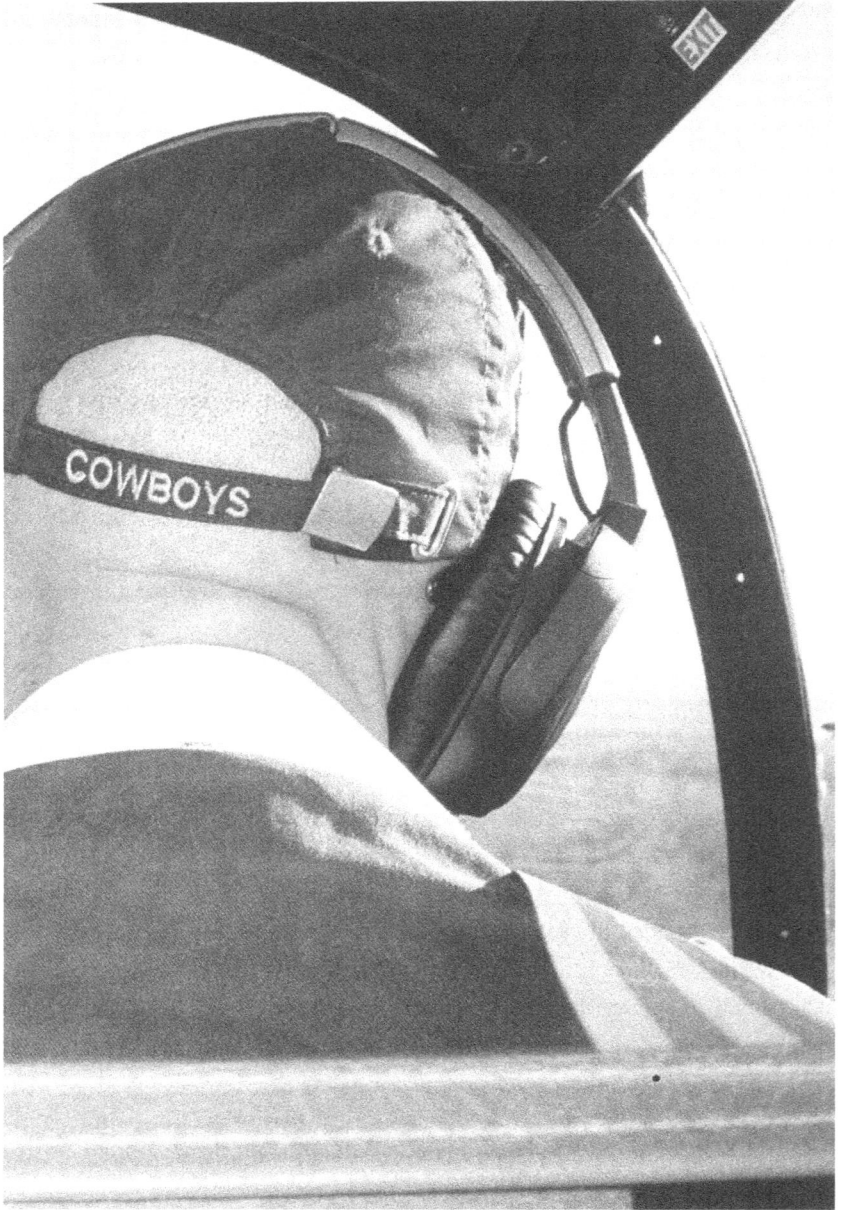

Helicopter pilot, Rinyirru National Park, 2018.

FOUR

Pigs

It was an early morning at Rinyirru National Park, and all the rangers had gathered under the mango tree for the usual morning meeting. Ranger in charge Ray was discussing the park's recent purchase of a new state-of-the- art pig trap. Ray explained that this particular pig trap was designed so that pigs would be lured inside the small enclosure by a carcass placed there, and would be free to come and go for some time. The enclosure was fitted with a motion sensor camera, accessible via a mobile phone or computer, that switched from still images to video when it sensed movement, and a remotely operated gate. The idea was that once there was a significant number of pigs in the trap at one time, whoever was accessing the camera could remotely shut the gate, trapping the pigs inside to be culled later. Ray said that he was hopeful that the pig trap would help to make their pig control more successful. However, at this stage the pig trap had not yet caught anything. Ray posited that this may be because the bait inside the trap (a dead cow) had not yet decomposed enough to really appeal to the pigs. He said that he might visit the pig trap later in the day to "poke a few holes" in the carcass to help it "ooze."

A few hours later, I was at the pig trap with Ray and another ranger, Roger, helping to secure tape around Ray's wrists so that the skin between his latex gloves and hazmat suit was adequately covered. As well as these, he wore knee-high rubber boots, goggles, and a mask. He told me that he was being cautious, in case the carcass was filled with gases and ready to explode all over him. Roger and I observed at a distance as Ray advanced into the pen of the pig trap, star picket in hand. Ray carefully approached the cow carcass, which by now was emitting quite a strong odor, and raised the star picket over his shoulder. He javelined the star picket into the cow's belly but the cow did not explode. Ray yanked the star picket out of the carcass and stabbed it a few more times for good measure. As we watched, Roger told me quietly that he was unsure if Ray's attempt to manipulate the bait in this way would have much of an effect. He explained that he thought the trap's proximity to a lagoon was potentially what was leading to the lack of success with the pig trap. " 'There's a big croc in that lagoon," Roger told me. "That could be keeping the pigs away."

Methods of Control

This pig trap was the latest addition to a multifaceted pig control program administered by the Rinyirru National Park rangers and Queensland Parks' specialist pest-management team. Feral pigs (*Sus scrofa*) are a strong focus for pest control in Rinyirru National Park. Pigs are exceedingly common and can be incredibly destructive. Some in the scientific community consider them to be "one of the most destructive invasive species in the world" (Negus et al. 2019, 581; McClure et al. 2018). In Cape York, feral pigs seep into places, degrading and disturbing wetlands through their feeding behaviors, which include rooting, digging, pugging, and wallowing. These behaviors result in the destruction of vegetation, erosion, and the compacting of moist mud. Feral pigs damage aquatic and shoreline vegetation, impact epigeic invertebrate populations, and alter the water quality through their urination and defecation (Marshall et al. 2020). They also feed on a range of vulnerable species, including water lilies and turtle eggs (Negus et al. 2019).

Like weed control, managing pigs involves killing to enact care. Here, killing takes on a different texture; it is more direct, more hands-on, more absolute, and is evaluated based on impacts. As with weeds, land managers kill pigs to enact care toward sensitive ecosystems, places, and other species. However, in this chapter I am interested in digging into the contradictions and diverse valences that emerge around who is allowed to be killing as care, and whose killing is seen as uncaring and problematic. Forms of authority structure how the killing of pigs is framed— sometimes as conservation, sometimes as antisocial and illegal behavior.

While Cattelino (2017, 131) points out that what constitutes a "harm" effected by an invasive species differs between different people, pigs are accepted by all land managers in Cape York to be a threat worthy of control and containment. The types of pig-induced harms cited by various land managers differ, related to each manager's livelihood, values, and concerns. As with weeds, land managers' pig control is about preserving particular kinds of preferred landscapes that allow for particular kinds of flourishing.

Aboriginal ranger groups and Queensland Parks seek to control pigs because of the damage they cause to sensitive wetlands, whereas graziers are concerned about ramifications for cattle management. Wetlands tend to support populations of water lilies, water birds and other animals, and are often significant story-places to Aboriginal Traditional Owners. The rangers and pest-management team employ a number of different techniques to control pig populations, including baiting with 1080 poison, aerial shooting, and nighttime shooting on the ground aided by an infrared scope. As ranger Ray told me, pigs are highly intelligent and many (particularly the older boars) know to hide when they hear the sound of the helicopter, which makes them difficult to shoot from the air. Ray explained to me that nighttime shooting appears to be ineffectual if one only considers the numbers, but if the shooter is able to target and kill a small number of problematic pigs—like those older boars—it can still be useful.

One of the specialist pest-management rangers explained to me that there are a few issues with baiting. To target introduced animal species, 1080 poison (sodium fluoroacetate) is used widely across Australia and Aotearoa (New Zealand). It is a useful poison in Australia, because the acid form of fluoroacetate is known to exist in around thirty species of native Australian plants (Department of Agriculture and Fisheries 2017, 1). This means that many native animals in Australia are not as susceptible to the poison as are introduced species, and, if administered correctly, the effects of 1080 on nontarget populations can be minimized

(NSW Threatened Species Scientific Committee 2019). In Aotearoa, activists have argued that 1080 has devastating effects on bird populations, resulting in forests that resound with silence—claims that are disputed by environmental managers and scientists (Addison 2022). In her work on 1080 use in Aotearoa, Courtney Addison suggests that the poison "becomes a pharmakon: literal poison and putative remedy, agent of life through death, morally ambivalent but ostensibly indispensable" (2022, 25). Through the death of some invasive life forms, 1080 allows other forms of life to flourish. It is what Addison calls "a compromised cure" (2022, 30). The poison is controversial not just because of its impact on nontarget species, but because of how it works. Royal Society for the Prevention of Cruelty to Animals scientist Sherley (2007, 456) has argued that 1080 is not a humane poison as it results in a prolonged and painful death to the target species. As such, Sherley argues that alternative methods should be considered. The 1080 toxin can take between 1.9 and 47.3 hours to begin working for pigs, with death occurring between 2.8 and 80.0 hours after ingestion (Meurk 2015, 97).

Although some animals present in Australia since before colonization are not susceptible to the 1080 toxin, dingoes are. The poison is distributed through lumps of meat that have been injected with the toxin. The pest-management ranger told me that in the past, they had used aerial baiting in Rinyirru National Park, dropping the poison-laced meat from a helicopter around the park. This was effective because they could cover a large area quickly and access places that are not reached easily on the ground. Now though, this ranger explained, they can only bait on the ground in specific areas. This is because of the potential for the baits to kill dingoes. As well as being a protected species inside of National Parks in Queensland, dingoes are a totem for some of the Aboriginal clan groups that are Traditional Owners for the park, and the accidental death of dingoes is unacceptable to them.

The cessation of aerial baiting in Rinyirru National Park has also been frustrating to neighboring graziers, who have noted an increase in pig numbers since the pig control regime in the park changed in accordance with Traditional Owners' wishes. Graziers spend significant time and resources on pig control. In Cape York, the average agricultural property spends more than AUD$9200 (USD$5943) each year on feral animal control, with most of this directed at controlling pigs (Marshall et al. 2020, 2209). Grazier Alan told me that he believes pig management needs to involve a cohesive effort from all neighbors, as pigs do not respect land tenure boundaries. Pig control tends to be everyone's business in Cape York. For graziers, the most significant issue with pigs is that they churn up the soft mud at the edge of waterholes and dams, transforming these areas into boggy wallows that can be dangerous for cattle, particularly as they weaken toward the end of the dry season. As grazier Bill explained to me, "we've got man made waters, dams and you get a mob of pigs hit a dam and root up around the edge, you get an old skinny cow going in there, next thing, bogged, you know." Bill's wife, Diane, nodding, told me, "and we get 'em good, cleaned out. [But] when it gets really dry everywhere else they all come for our dams."

Graziers use a number of techniques to control pigs. They are assisted by the Cook Shire Council to bait for pigs using 1080 injected into lumps of meat. Graziers can apply to take part in this project, and a practitioner will visit their station and supply the materials required. Where these pest projects were previously administered by a nongovernment feral pest and weeds organization affiliated with Cape York Natural Resource Management, legislative and regulatory changes around what constitutes an "authorized person" to distribute 1080 have functionally meant that only government organizations can administer 1080 in the area. Many graziers complained about this change, arguing that this made it more difficult for them to access the project and, accordingly, more difficult to control the populations of pigs on their leases. On occasion, CYNRM will offer to conduct aerial shooting on graziers' leases, but many graziers prefer to shoot pigs themselves. During a visit to one cattle station, just as we were sitting down to the evening meal, grazier Alan said that he had seen some pigs just minutes before as he moved some cattle from a lane into the yards. His son—eighteen years old and very interested in shooting pigs—leaped into action, grabbing a bike and a gun, and riding off into the night. He was absent for some time, presumably stalking the pigs, while the rest of us chatted and ate. After a while, the sound of shots reverberated. The son returned, triumphant, announcing that he had successfully shot three pigs but that some (including the biggest boar) had managed to get away. As well as shooting, many graziers used yellow phosphorous, which they referred to as "sap," to control pigs. Sap is painted onto animal carcasses for pigs to consume. It is less publicized, less controversial, and less regulated than 1080 (Stephan 2006, 15), and as such is the preferred poison for graziers as it enables them to carry out their pig control without requiring a government-approved practitioner to be present.

Singh and Davé (2015) suggest that killing animals draws humans into some kind of ethical engagement. In asking, "what is it to kill, to kill well, to be killed or killable?," they argue that considering the human role in nonhuman death is a question for the anthropology of ethics (Singh and Davé 2015, 245). They point out that the act of killing, or rendering an animal as "killable," provokes an ethical relationship, or quandary, between humans and animals in which the executioner is, in some sense, morally culpable. They suggest that the kind of companionable relationships between humans and animals that scholars like Haraway (2008) describe and theorize do not necessarily provide a solution to these kinds of ethical quandaries. Singh and Davé distinguish two types of animal killing: the sacred and the profane. Sacred killing of animals is killing done for ritual or sacrifice, whereas profane killing refers to killing for subsistence or commercial activities. Their use of the term "killable" is what I wish to draw attention to. Of "killability" they write: "if the thing is killable, is the act experienced as killing by the executioners? How does killability shade into vitality, and how does it not?" (Singh and Davé 2015, 233).

Pigs in Cape York are considered virtually across the board as "killable." This killing is neither sacred nor profane, like the forms of killing described by Singh and Davé, although some pigs are indeed killed for subsistence. Pigs in Cape York are considered "killable" because they fall into a third category of killing that Singh and Davé do not go into: culling. In some instances, the killing of pigs alongside weed control can be considered part of a purity project. This is particularly true of Queensland Parks, which as an institution seeks to return Cape York to a preinvasion "pure" site of natural nativeness, and in its ideological bent sways close to a kind of "sacred" killing without really getting there. For the killing of pigs to be characterized as culling, eradication, or control, pigs undergo a reclassification from nonnative, to invasive, and finally, to pest. Considering pigs as "pests" deems them not only killable but also attributes to them a negative agency (Einarsson 1993, 74; Robinson, Smyth, and Whitehead 2005, 1389). The language of culling minimizes some of the emotion evoked by the idea of killing, somehow more abstract, scientific, and removed. Culling becomes a technical matter rather than evoking the intimacy that killing necessarily entails. While obscured by the scientific discourse of environmental management, pigs are characterized by—in particular—Queensland Parks rangers and graziers as willfully causing damage and exhibiting deviant behaviors, when in reality pigs are behaving exactly as they ought to, fulfilling a biological destiny to dig, root, wallow, and consume. Pigs are considered to be particularly deviant when they go about their business on cattle stations and in national parks.

Having already been deemed "killable" by environment management discourse, legislation, and the mandate of Queensland Parks, the act of killing is further abstracted and removed from individual culpability by the use of 1080 baiting (Meurk 2015, 98). With 1080 baits, in particular, the decision to kill the pigs has already happened, way up the chain of command and somewhere physically distant from where the killing takes place, in a government office in Cairns, Brisbane, or Canberra. The ranger who checks whether or not the poison baits have been consumed is unlikely to be the same ranger who laced and laid the baits, and certainly is not the same ranger who authorized the killing. These layers of distance transform the death of the pig into something abstract, working to erase individual culpability and any form of intimacy between the killer and the pig. The act of killing, then, is reduced to scientific abstraction—a technical matter—and the violence of killing is obscured. Recall how Marvin, who works in feral animal management for a nongovernment organization, spoke to me of being chased by bullocks in his dreams after the mass cull of cattle in the 1970s tuberculosis scare. No one spoke to me about being similarly haunted by pigs, gesturing at the different ways in which these two introduced species are valued. Yet as Meurk (2015; 2011) and Sherley (2007) point out, the use of 1080 toxin is not a painless or necessarily a humane death.

Killing as Care

I contend that we can understand killing pigs—particularly for explicitly environmental management purposes—to be an act of care. Aside from any notions we may have about pigs exhibiting deviant behavior or being ascribed negative agency, there is the undisputable fact that their presence in Cape York's fragile wetland ecosystems wreaks enormous amounts of damage.

Land managers in Cape York today have inherited a landscape already damaged by the impacts of colonization, extractivism, and poor management practices. They have inherited a landscape already rife with invasive species, like pigs. In encountering this inherited landscape, day by day, there is an ethical demand placed on land managers to respond. A "willingness to respond," is, according to Martin, Myers, and Viseu (2015, 634) a necessary component of care. Cattle graziers, Queensland Parks rangers, and Aboriginal Traditional Owners respond to pigs in various ways, but common among all land managers is the taking part in a difficult relation of care. All land managers are, to varying extents, involved in culling pigs.

These land managers are doing the messy work of "caring for," not just "caring about" (Puig de la Bellacasa 2017, 4–5), when they take part in pig control. Such an act of care emerges from an embeddedness in the land and region; it is interested and noninnocent. To be clear, the care being enacted here is not toward the pigs themselves. Similar to the case of the goat eradication in the Galapagos islands, which Bocci (2017, 437) recounts in engrossing detail, the way that care plays into the killing of pigs for park rangers is less about ensuring a "good death" for the pigs and more about achieving maximum killing. Nor is this form of care analogous to the ethical refusal that Blanchette (2020, 116) details among some workers tasked with the artificial insemination of pigs in industrial pig farming in the American Midwest, in which ambivalence, distancing, and refusal can be read as a kind of care. In culling pigs in Cape York, though, care is directed toward the health of the ecosystem and biodiversity of the region as a whole—the sensitive wetlands, estuaries, and water lilies that Cape York is renowned for.

The notion of relations of care involving violence is not new. As van Dooren writes, "in this time of extinctions, it seems that care and hope are frequently saturated with, perhaps even grounded in, unavoidable and ongoing practices of violence" (2014, 91–92). In work by a range of scholars, including van Dooren (2014; 2019), Wanderer (2020), and Bocci (2017), the concept of care has been used to think about the control of invasive species. Many of these scholars speak about how, in culling an invasive species in order to protect a vulnerable one, a relation of care is enacted. This care is generally for the endangered or vulnerable species, but often a form of care is extended toward the species targeted for control.

Bocci (2017) talks about how in the explicitly violent relationship between state-employed hunters and goats in the Galapagos islands, the hunters' work required them to become attuned to the goats. This cultivation of the "art of noticing" (Tsing 2015) is an expression of care, even if the observations garnered were used to ensure that culling became more effective, more absolute. Wanderer (2020), similarly, discusses the attunement generated among scientists toward the rodents they research—not yet the target of culling, but soon to be so once enough data has been gathered to inform an effective plan. In his work on the conservation of the endangered whooping crane, van Dooren (2014, 92) discusses how a "regime of violent care" is enacted, with some "feathered bodies" sacrificed and used in the service of others. As he writes, "in the context of conservation biology and its material practices, it is clear that care for the species often trumps other considerations, including the well-being of individual animals" (van Dooren 2014, 108).

For cattle graziers, pigs are killed to enable the flourishing of another introduced, "exotic" species: cattle. The bogginess that pigs create at the edges of waterways and dams can become deadly for cattle toward the end of the dry season as grasses dry out and cattle, in particular the older females, begin to weaken. It is around this time that graziers tend to monitor their cattle more closely. They drive around their stations, delivering nutritional supplements to the cattle to help them digest the poor-quality grasses that remain and checking on the quantity of water in their dams. When I accompanied graziers on these journeys (bouncing along on the cracked vinyl seats of various old Toyotas, sweating in the hot sun), I was more often than not sharing my seat with a propped-up old rifle. The rifle was there in case of two eventualities: in case the grazier saw a pig, or in case the grazier came across a bogged cow. The gun, then, could serve to both euthanize the loved cow or cull the unloved pig. On these outings I never came across a stuck cow, and only once witnessed a grazier fumble with his gun, shoot at, and miss a number of pigs by a waterhole, succeeding only in dispersing them.

For park rangers, killing pigs is enacting a form of care for biodiversity and sensitive ecosystems, rather than any one species in particular. Pigs are understood to be causing widespread damage and are targeted using a variety of techniques: shooting, trapping, and baiting. However, within national parks, care is also directed at preserving and protecting parks as a particular type of space; somewhere that has the visual amenity of a renowned "wilderness" space. Pigs are animals out of place in the so-called wilderness of Cape York. They disrupt and disturb the expectations that tourists may have about encountering a certain type of native and "natural" landscape. The damage they wreak on wetlands is visually confronting—long stretches of pocked and boggy mud that is difficult for other species to traverse, large wallows, decimated vegetation. But is this care for the ecosystems, landscapes, and biodiversity of Cape York undone when Queensland Parks works so hard to punish Cape York locals and visitors for pig hunting? Does this undermine the killing as care that happens in pig control?

Despite their classification as "killable pests" and despite a concerted effort by a variety of land managers to reduce pig numbers, pigs remain an intractable part of the Cape York landscape. As with weeds, there is a resignation among land managers that it is unlikely that Cape York will ever be truly free of pigs, but the desire to control pigs is related to the valuation of particular types of landscapes. Among Aboriginal Traditional Owners, pigs occupy a more contingent and contextual space in which they exist not only as "killable pests," but also variously as an economic resource, type of pet, and source of recreation. In this sense, pigs emerge as disrupting an assumed correlation between "nativeness" and "belonging." Where pigs are controlled, Aboriginal Traditional Owners are enacting care not only for biodiversity, ecosystems, and other species, but also for the ancestral spirits that dwell within their homelands.

The Ambiguity of the Invasive

On an afternoon in the early dry season, I visited the small coastal outstation community of Yintjingga with Lama Lama rangers Leena, Mabel, and Jackson. We stopped in at Leena's grandparent's house, located in a small group of houses surrounded by mango trees known as Bottom Camp. A litter of mixed breed pups was trotting around, following Leena like a shadow as she quickly checked the house and yard for any sign of her grandparents. A pot was boiling over on an outdoor fire near the house, a pig's trotter just visible above the bubbling water. I wandered over to the pot for a closer look, and Leena explained that it was wild pig cooking. She told me that she was excited at the prospect of eating it later in the evening. We left Bottom Camp and found Leena's grandfather, Harry, at the beach, along with other relatives. On the short drive to the beach, Jackson told me that the bush provided them most of the food they needed to survive. He reeled off a list of bush and marine resources: fish, plains turkeys, stingrays, turtles, dugongs, and pigs.

Later in the evening, we drove Leena's nephew and Harry back to the house. We were just dropping in, needing to return to where the rangers were camped for a couple of weeks of work about an hour's drive away. Harry portioned up some pig for us to eat before our drive. He explained that a relative had accidentally hit the pig with her car and brought the pig to Harry to cook. Harry told me that he had put the pig in a pot of salt and water on the fire and cooked it all day. While settler-descended residents of the region tend not to eat wild pig unless it has been hand-raised, finding the flavor to be quite strong and gamey, eating wild pig is a fairly commonplace and much-enjoyed occurrence for Aboriginal people in Cape York. When I ate the pig, careful to avoid a few rogue hairs that survived the boiling pot, I found it to be pleasantly flavorsome and very tender from Harry's slow cooking. When we returned to the rest of the rangers back at camp, Leena happily announced that we had just eaten pig, sparking expressions of jealousy and annoyance from the other rangers, disappointed that we had not brought meat to share.

Pigs are widely considered by Lama Lama and other Aboriginal people in Cape York to be an important bush resource, despite their status as an introduced species. Jackson made no distinction between those bush and marine resources that are native animals and pigs, describing them all as important parts of the diet of Lama Lama people who live at Yintjingga. On another occasion, as I drove around Rinyirru National Park with Kuku Thaypan ranger Donna, she asked me if pigs were an introduced or native species. When I said that they were introduced, she told me that this is what she had thought, but she wanted to confirm. Her lack of clarity makes sense given the length of time that pigs have been present in Cape York, and the important role they play as a food staple for Aboriginal people of the region. Several people in Cape York—including some park rangers—referred to the wild pig populations as "Captain Cookers," indicating a belief that these pigs are descended from the several pigs that Captain Cook released at the site of present-day Cooktown in 1770. If this was, indeed, the origin of pigs in Cape York, this would mean that Aboriginal people had been living in relationship to pigs for a significant amount of time before the region was colonized by European miners and graziers. However, the consensus among the scientific community of Australia is that it is unlikely that pigs arrived in Cape York in this way. Much more likely is the widely accepted explanation that feral pigs in Cape York are descended from escaped domestic stock farther south. This is based on the fact that the earliest reports of feral pig sightings in the region occurred in European explorers' journals in 1847 (Meurk 2015, 88). And yet, the myth of Cape York pigs' origin demonstrates how embedded they have become in the social and ecological worlds of Cape York.

The ambiguity of pigs is related to a broader slipperiness around the native/introduced binary. The concept of "nativeness" relies on an arbitrary temporal boundary. In Australia, this boundary is set at 1788, the date of colonization. Such a boundary acknowledges neither the impact that Indigenous people precolonization had on the distribution of species, nor the differential and varied ways in which colonization happened (and continues to happen) in Australia. In addition, if we consider the threshold of "nativeness" to be 1788, what does this mean for the pigs in Cape York if they really are descended from those let loose almost two decades earlier?

In Helmreich's (2005) work on how scientists think about native species in Hawai'i, what counts as native or indigenous to Hawai'i depends on who is talking. This aligns with Head's point that the category of "nativeness" is related more to human bounding practices than any kind of biological or evolutionary reality (2012, 171). According to Helmreich, Native Hawaiian people categorize species like taro to be indigenous to Hawai'i even though taro was brought during the original human settlement of the islands. Despite the human role in the distribution of taro, it is a species that is understood as indigenous by Native Hawaiian people and functions as an important symbol for Hawaiian sovereignty movements (Helmreich 2005, 111).

In Cape York and even Australia more broadly, a comparable species is the dingo, which the archaeological record suggests has been present in Australia for between 3,000 and 5,000 years (Crowther et al. 2014). Dingoes are widely considered to be native animals in Australia, which makes sense if we take European colonization to be the temporal threshold of nativeness (Head 2017). While broadly understood as native, dingoes are only protected as a native species inside protected areas like national parks (Queensland Government 2016). Dingoes are a complex species to consider because they can breed with domestic canines, and there remains debate within the natural sciences as to whether these hybrid offspring ought to be categorized as dingoes or wild dogs (Crowther et al. 2014). This complicates any efforts toward conserving or protecting dingoes (Daniels and Corbett 2003). Graziers and other land managers outside of national parks frequently engage in "wild dog" control, a category that functionally includes dingoes. Not only are they within the law when they do so, but under the Biosecurity Act 2014 landholders are required to take "reasonable steps" to control wild dogs (Queensland Government 2016). Yet to the wider Australian public and to Aboriginal people who in various parts of the country relate to dingoes as totems, dingoes are understood as native. Dingoes, then, represent another kind of species that counts as native or not—depending on who is doing the categorizing.

For Aboriginal people, characterizing particular species as "belonging" in Cape York does not always correlate with nativeness. Cattle are understood variously as a species that belongs and indeed makes life viable in Cape York for Aboriginal Traditional Owners alongside graziers. While cattle are never mistaken for native species, they are considered by graziers and some Aboriginal Traditional Owners to belong in the region in some sense. Pigs, too, are generally understood to be introduced, but they occupy an ambiguous position for Aboriginal people—sometimes belonging, sometimes in need of exclusion and control. Lama Lama people understand that pigs need to be excluded from specific, sensitive areas like fragile wetland ecosystems and story places. At a planning meeting involving the Lama Lama Land Trust and Queensland Parks, feral pigs and cattle were discussed as a threat to Lama Lama National Park. One young Lama Lama ranger spoke about how both feral pigs and cattle are "destroying the place," and all the rangers gathered agreed, saying that cattle and pigs have an impact through their eating, digging, trampling, rooting, and camping (sleeping). In particular, the Lama Lama rangers pointed out the damage that pigs cause to significant cultural sites. There was a consensus that pig damage ought to be addressed through aerial shooting and baiting of pigs as well as the fencing of sensitive and culturally significant places.

I accompanied some Lama Lama rangers on an expedition to fix the pig-proof fencing around a wetland in the early dry season. I set off with two of the younger rangers—Essie and Jackson—to inspect the fence. Our job was to look for places in the fence where it looked like pigs had been getting in, or where the fence had been damaged. We were equipped with bright pink tape, which we tied around the fence where it needed attention. A team with fencing gear was following us, riding an all-terrain vehicle called a Polaris around the swampy perimeter of the wetland. The wetland was dotted with water lilies and partially submerged melaleuca trees. We spotted a blue-winged kookaburra at one point, a rasp snake darting through the shallow water at another moment. There was a flurry of excitement when the team working on the fence repairs startled an adult pig and a throng of piglets. Much to the delight of his coworkers, a young ranger named Wayne dove for— and successfully caught—a tiny black piglet. He proudly presented it to Jackson, who had been hoping all week to catch a piglet that he could keep as a pet. Jackson and Essie were overjoyed, kissing and cuddling the squirming piglet. We soon discovered that the little creature was covered in ticks, which Essie, Jackson, and I dutifully spent the drive back to camp gradually removing. I asked Jackson if he planned to eat the piglet when it grew bigger. He said that he would spay it and keep it as a pet. Back at camp, the boys constructed a temporary pen for the piglet out of tables, chairs, and boxes. For several days, Jackson's leisure time was oriented around the little piglet. Before long (and seemingly inevitably), the piglet escaped.

Pigs, along with other introduced animals and plants, can be read as unintentional nonhuman colonizers. Like human colonizers, these species have spread into new landscapes, altering ecosystems and both enabling and imposing new relationships with the humans and nonhumans who already lived there (Tsing, Mathews, and Bubandt 2019). However, unlike cattle, pigs were not intended by settlers to aid in outnumbering

Water lily, Rinyirru National Park, 2019.

Lama Lama rangers overlooking Bull Swamp, Lama Lama National Park, 2018.

Aboriginal people in order to take control of land. Instead, pigs were brought to Australia, where they have no natural predators, and escaped the bounds of how they were originally intended by humans to behave here, along with other invasive species.

To Aboriginal people, pigs are not simply a pest animal to be controlled. Instead, they are an important food resource that also has the potential to be destructive to sensitive areas, and ought to be excluded from these areas—although none of the Aboriginal Traditional Owners I worked with were opposed to culling pigs, and pig control did not draw the same level of contention and concern as cattle management. Pigs, then, are considered by Aboriginal people to contingently "belong" on some level and in some places in Cape York.

Such incorporation of an introduced species by Aboriginal people is not unique to pigs in Cape York. Other authors (see, for e.g., Seton and Bradley 2004; Robinson, Smyth, and Whitehead 2005; Trigger 2008; Trigger 2012; Martin and Trigger 2015) have noted similar instances across Australia in which introduced species such as donkeys, camels, and horses are sometimes incorporated into cosmologies, and sometimes—as in Cape York—in ways that are mundane and practical. As Vaarzon-Moral (2017, 188) has pointed out, "Aboriginal peoples' responses to exotic animals cannot be characterized simply in static, dichotomous terms of resistance-acceptance and belonging-not belonging." Instead, these relations entail more ambivalence; they are contextual and iterative, they change over time and vary between individuals even within the same locality. Many Aboriginal people have relationships to introduced species that are contingent and contextual. Pigs are simultaneously a valuable food source, valued pets, and a threat to sensitive ecosystems, vegetation, and culturally significant sites. Aboriginal Traditional Owners' work to cull and control pigs can be read as an explicit act of care—for landscapes, biodiversity, ancestral spirits, and for culturally significant sites, even as they live alongside pigs. It is the possibility that pigs can damage specific valued places and threaten other valued species that leads to their censure and control, rather than a conviction that pigs need to be removed from the region entirely.

Pig Hunters

On a hot day, toward the end of the dry season, I found myself stranded at the southern ranger base in Rinyirru National Park. As would happen, from time to time, one of the rangers from the northern base had dropped me off at the southern base so that I could try to talk to the rangers located there while she attended to some other tasks. She told me that she would collect me in an hour or two. I wandered around, soon coming to sit in the shade of a large shed, hopeful for an afternoon breeze. Before long, a ranger called Shane drove into the base and pulled up near me. He asked me if I would like to sit inside his air-conditioned office, a small prefab building located on the northern side of the base.

The interior of the prefab was, like many Parks offices, fairly rudimentary. Shane had a desk, a computer, a couple of well-used office chairs, and a filing cabinet. The walls were lined with laminated Parks materials, mostly maps and safety information. Some faded tourism pamphlets were scattered across his desk. Unlike the offices in the northern ranger base, Shane's office was not open to the public. Shane sat down, placing a small, tough plastic camouflaged-pattern box on the table. I recognized this as a motion-sensor camera, designed to be hidden in trees. Such cameras are used for a variety of purposes. Many conservation biologists and ecologists use these kinds of cameras to monitor both pest and native species. Hunters, too, use these cameras to get a sense of where particular animals spend time. Shane, however, used these cameras as a surveillance tool in order to monitor humans. Specifically, Shane was seeking to catch pig hunters trespassing on the park and transgressing Parks rules. Shane told me that he had seven or eight of these motion sensor cameras hidden around the park, on trails that are not supposed to be accessed by the public.

"I don't tell anyone where the cameras are," he told me, explaining that his hidden cameras often catch relatives of Parks employees, or employees themselves. "I don't trust anyone to not do the wrong thing from time to time."

Shane pried open the camera's protective casing and removed an SD card, which he inserted into the aged government-issued computer. Sitting at his side, I went through the images with him that the camera had captured. Most were slightly blurred photographs of four-wheel-drive vehicles, often with one or two people sitting on the cargo rack. Shane explained to me that the previous weekend a pig-hunting competition had taken place in the nearby town of Cooktown, and that this annual event generally resulted in a number of people coming into the park illegally to hunt pigs. Shane was looking for legible license plates that he could record. Together we peered at the grainy images, trying to distinguish letters and numbers. Some license plates were easy to read, others were obscured with mud. Shane pointed this out, saying that many people do this on purpose as they are aware of the presence of hidden cameras, and of the fines that they will receive if caught hunting within the park boundary.

I asked Shane how he proceeded when he had a license plate recorded. He told me that he passed the information on to the Queensland Parks compliance officer for the region, an ex-police officer who traveled between remote area parks issuing fines and infringement notices. This compliance officer, along with senior on- and off-park management, would decide how to proceed on a case-by-case basis. Shane explained that in a place like Cape York, issuing fines and notices can be sensitive. Especially with the Cooktown pig-hunting competition, many of the hunters being fined were locals, and often these locals were relatives of Parks staff. As such, it was easier for Shane to not be directly involved in compliance infringements. Because of the floating nature of the role, the compliance officer did not face the same interpersonal complications in issuing fines and notices.

Pig hunting and pig hunters tend to be characterized negatively by Queensland Parks rangers, despite the fact that some rangers do pig-hunt recreationally. On her work on pig control in the Douglas shire region of far north Queensland, Carla Meurk (2015; 2011) found that some scientists and environmental managers characterized pig hunting as a debased form of recreation that glorified violence and killing. Pig hunters frequently use pig dogs, large mixed-breed canines that track and chase pigs, holding them down by the ears so that the human hunter can approach the pig and kill it—usually using a knife that is stabbed through the pig's heart. In his work on Australian pig hunters, Paul Keil states that, "pigdogging methods are intimate and violent" (2021, 100), and that such methods—in their spectacle, in their goriness—invite sensationalized characterizations of pig hunters by the general public.

While Meurk found that not all environmental managers took such a moralizing tone, most still tended to categorize hunting as "illegitimate killing," whereas scientifically ordained pig management (trapping and baiting) was deemed "legitimate killing" (2015, 93). These environmental managers draw a distinction between killing pigs for management (culling), a process intended to enhance life, and killing pigs for sport, which some perceive as glorifying death (Meurk 2015, 93). In this way, the environmental managers Meurk worked with positioned themselves as having "appropriate intentions" in contrast to hunters. Meurk highlighted some of the slipperiness around this, though, when she asked an environmental manager who had engaged in spotlight shooting whether he had enjoyed the task, and he replied that he had (2015, 93).

In Cape York, though, it is not a denigration of the way that pig hunters kill pigs that provokes Queensland Parks rangers to characterize pig hunters negatively. Rather, it is the flouting of the rules around park use that is the focus. The focus on controlling and censuring pig hunters, as well as pigs, is a central tension in Queensland Parks' pest control program. Such a situation brings to the fore questions around whether Queensland Parks is more concerned with achieving conservation outcomes or enforcing rules and regulations.

In Radhika Govindrajan's (2018) work among villagers in Kumaon in the central Himalayas, wild pig control was a contentious issue. Here, the state designated pigs as "wild," and thus deserving of protection from (in particular, low caste) humans. Villagers, though, sought to control pigs who they deemed as both dangerous and a threat to their subsistence and livelihoods, given that pigs tended to decimate crops. Villagers were highly restricted in their ability to control pigs, facing repercussions if they were caught killing these cared-for animals. In this situation, the villagers, too, were coded as "wild," but with wildness ascribed negatively. It was the unruliness of these villagers that the state sought to control, rather than the pigs themselves, even as these pigs terrorized villagers and consumed the food they relied on. For Govindrajan, wildness here functioned to simultaneously designate particular actors as threatening and under threat.

In Cape York, both pigs and pig hunters are framed as threatening, as unruly, and as in need of control. Pigs and pig hunters are simultaneously understood as a threat to the landscape and imaginaries of Cape

York, disrupting the fiction of a pristine, naturally "native" place. Both emerge as unruly, as out of place. And yet, some of these pig hunters are likely Traditional Owners for the areas on which they hunt. As such, their actions could be read as Traditional Owners exercising their Native Title rights to access resources from their homelands. This is not how the situation is read by Queensland Parks rangers or compliance officers.

While Aboriginal Traditional Owners are technically allowed to hunt in the park, this is a right that in practice does not seem to extend beyond fishing. In conversation with the ranger in charge for the park, it became clear that his interpretation of the rights of Aboriginal Traditional Owners to access resources in the park was limited. He told me that, for fishing, Traditional Owners must follow the same rules as everyone else, and that Traditional Owners can hunt in the park provided that they use "traditional methods." While he did not go into detail about what such "traditional methods" could be, it is clear that pig hunting (which tends to involve utility vehicles, dogs, guns, and knives) was not considered by Queensland Parks staff to be an exercise of Traditional Owners' Native Title rights.

The distinction between "traditional" (and, thus, state-sanctioned) and "untraditional" (and accordingly unsanctioned and warranting control and censure) ways in which Indigenous Peoples access resources has been heavily critiqued by anthropologists and other scholars, as it relies on the violent "noble savage" trope (Rowland 2004). In this, indigenous people are situated as being either authentic, premodern, and frozen in a temporally distant precolonization past, or as corrupted by modernity, having experienced cultural loss and, accordingly, as being less authentic, less culturally distinct from settler populations, and somehow less indigenous. In such reifications indigeneity is positioned as "that which modernity lacks" (Braun 2002, 91). There is no space, here, for indigenous cultures to do what all cultures do: change, transform, and adapt. As Braun writes, "at the point that indigenous peoples become too closely associated with the social, economic, and technological relations that signify the modern, they lose their claim to indigeneity" (2002, 94).

When it comes to hunting, conservationists often cloak these framings in the language of sustainability. In discussing the ways in which conservationists have framed Aboriginal people in Australia, Sackett (1991, 242–43) recounts how a particular environmental scientist from

Sydney feared that "the introduction of motor boats and fishing nets and four-wheel drive vehicles and rifles had led to the Aborigines [*sic*] achieving a distinct upper hand," which may result in the overexploitation of species. This kind of rhetoric is echoed around the world, with very real impacts on the ability of Indigenous people in settler-colonial states to access resources. In situations where the species being hunted is protected or vulnerable, such rhetoric contributes to conflicts in which Indigenous peoples are positioned as environmentally destructive by conservationists or animal rights activists (Beldo 2019; Arnaquq-Baril 2016; Nursey-Bray, Marsh, and Ross 2010). Yet, here, the species being hunted is a pest species already targeted for control. What, then, is driving the censure of pig hunting?

Within Rinyirru National Park and Lama Lama National Park, rangers and many Aboriginal Traditional Owners are concerned about pig hunters, regardless of whether they are Traditional Owners or not. Some of these concerns are tied into a broader care for the landscape, and it is these concerns that Traditional Owners tend to emphasize. Traditional Owners are worried that pig hunters will start fires that may spread out of control, that they may cut or damage fences, that they might drive their vehicles through the scrub instead of on designated tracks (something that has impacts in terms of both the spread of weeds and erosion), and that they might disrespect the Country and the ancestral spirits who dwell there.

At a planning meeting with the Lama Lama rangers, pig hunters were identified as posing a significant threat to sites that are both ecologically and culturally important. At this meeting, a Lama Lama ranger named Samson said that visitors to the park were often "pleasing themselves with the Country, going where they want to go, to good fishing spots and not realizing that it's a cultural site." He noted that not all visitors do the wrong thing, but that he had come into contact with "some dickheads" in his role as a ranger and traditional owner. Other Lama Lama people present expressed concerns about pig hunters entering culturally significant areas and, in particular, riding over ceremonial areas on quad bikes.

Outside of the national park, the relationship between Lama Lama rangers and pig hunters was more nuanced. On one occasion when I had visited Silver Plains to take part in a fencing project with the Lama

Lama rangers, a group of non-Indigenous young men visited to hunt pigs. These men were from around Townsville, about 900 kilometers (around 560 miles) south of Silver Plains, and had attended high school with Samson. Because of their relationship with Samson, these men were given permission to hunt for pigs on Silver Plains. Some of the older Lama Lama people present, including senior rangers and Lama Lama elder Charlie, expressed misgivings, suspicious that these pig hunters might do the wrong thing by lighting fires or shooting a cow in order to attract pigs. The younger Lama Lama rangers, however, did not seem to share these concerns. A couple of the younger Lama Lama rangers accompanied these pig hunters on at least one occasion at nighttime to spotlight for pigs, and the pig hunters gave a large pig that they had successfully killed to the Lama Lama rangers to eat, presumably as thanks for the opportunity to hunt on Lama Lama land and because they did not consume wild pig meat themselves.

Some of these concerns about damage to sensitive areas are shared by parks staff, but rangers tend to also be concerned with the impact that pig hunters may have on other visitors to the park. When I spoke with the ranger in charge of Rinyirru National Park about the impacts of pig hunters, he cited the illegality of their actions (bringing dogs and firearms into the park) and antisocial behavior (such as drinking or making noise in the campsites) as some of the key reasons that pig hunters need to be controlled and excluded from the park. These concerns align less with the care for the land and biodiversity that pig killing involves and instead are about facilitating tourism and visitor access to the park. In some ways, these concerns are less about care than they are about control. Perhaps, then, the preferred landscape of the park for Queensland Parks rangers is not only one that is "natural" but also one that is highly regulated. The park, here, is envisioned—primarily—as a site for tourism rather than as Aboriginal land or even land set aside for the purposes of conserving sensitive ecosystems and species. In describing the issues that pig hunters present, park rangers emphasize the importance of tourists having an experience that is not only positive, but anticipated. This echoes the concerns of the senior park rangers cited in chapter two about the presence of cattle disturbing tourists, and in chapter three regarding the spread of gamba grass in Nifold Plains. The control of pigs and pig hunters is a continuation of the broader project to present to tourists the park that they expect to encounter: a "natural," native landscape, devoid of pests and unruly, unsanitary locals and visitors. This is not to say that in seeking to control pigs and censure pig hunters park rangers are not enacting a relation of care toward the land, but that this care is bound up with and inseparable from a desire to create an orderly, expected landscape. These driving forces of care and control rub up against each other and produce tensions, particularly in this situation where the desire for control (a park without pig hunters) threatens to supersede the ethic of care (a park without pigs).

In his study of wild boar control in the suburbs of Barcelona, Arregui (2023) found that the individualized relationships that some locals had to particular boars tended to undermine efforts at removal. As one hunter told Arregui, "It doesn't matter what you do to fight the wild boar pest. In this city, there will always be someone who will help them to stay" (2023, 9). In Cape York, there is a sense among locals (park rangers and cattle graziers alike) that pig hunters like to leave some pigs alive to ensure that there will be a population to hunt in the future. Whether such a suspicion is based in fact or not is probably immaterial—the number of pigs being killed by occasional pig hunters pales in comparison to the number of pigs being baited, shot, and trapped by the land managers living and working in the region year-round.

Pigs continue to multiply in Cape York, but this is not because of any intimacy between particular humans and pigs that allows their continued dwelling. Pig populations in Cape York are entrenched. Pigs will continue to seek out the wetland ecosystems that they find attractive, and they will continue to reproduce. Because of their rate of reproduction, scientists estimate that each year 55–70 percent of the feral pig population would need to be killed to control pigs (Bengsen, West, and Krull 2018, 332). The vastness of the region and the relatively small human population residing there means that it is unlikely that pigs' numbers will be dramatically reduced any time soon. Some Cape York land managers see the reasons for pig proliferation to be more specific, traceable to particular policy decisions and land tenure changes. The increase of National Parks and Aboriginal land in the region, coupled with the insecure funding arrangements that employ Aboriginal rangers on these tracts of land in only intermittent fashion, mean that much of the land in Cape York is empty of people for much of the time. Graziers argue that when

Cape York was held primarily under pastoral lease instead of national parks, pig populations were kept under a greater degree of control. It is difficult to assess the validity of these claims, or even to accurately quantify the changes in pig numbers due to gaps in pig monitoring—both in the past and nowadays (Bengsen, West, and Krull 2018).

What is clear is that pigs, who may be better thought of as colonial remnants instead of foot soldiers for colonialism itself (Rose 2004), are likely to remain in the region. It is likely that land managers will need to continue to grapple with their impacts and continue to enact methods of control that are both violent and banal. Pigs—and pig hunters—are here to stay.

———

As Paolo Bocci (2017, 436) notes, eradication is "always rife with possibilities of failure." Actually achieving eradication seems to be virtually impossible. This is something of an open secret among land managers. Recall ranger Ivan from the previous chapter who joked, in relation to weed control, that "eradication is a dirty word." Pigs seem to find a way to linger, persist, and despite the best efforts of all land managers, proliferate. Cattelino (2017, 132) reminds us that "equilibrium is fantasy," and that the "natural" landscapes that land managers seek to protect and preserve are only made through ongoing caring labor and maintenance.

Interacting with invasive species, according to Cattelino, urges us to question what we mean when we talk about "diversity" (2017, 133). This brings us back to the point that it is important to engage critically with the idea of a "baseline" that we seek to return species and landscapes to (Trigger 2013). Do we assume that this baseline is based on the notion of a precolonial landscape, so far untouched by the spread of invasive species and the transformations inherent in agro-industry? Shotwell (2016, 4) reminds us that there is no pretoxic Eden that we can return to; no primordial landscape that can be recaptured. In a practical sense, even if returning landscapes to their precolonial status was the goal, gaps and failures in biodiversity monitoring—in the past, and today—make this an impossibility.

Land managers in Cape York are aware that they will never eliminate pigs from the region. They are not necessarily seeking to cultivate a pure, natural, and native landscape. What Aboriginal Traditional Owners,

Queensland Parks rangers, and cattle graziers are each seeking to do in their pig control is to cultivate and maintain *workable* landscapes. They are laboring in the hope of making do. Of course, what this means for each aggregate of people—and, likely, for each individual—differs.

Aboriginal Traditional Owners have multiple, overlapping concerns. They want to protect sensitive wetlands and endemic species; they want to maintain important cultural sites and exclude the damaging presence of pigs and pig hunters from these areas, at least in part to express care and respect for the land and for the ancestral spirits who dwell there; and many people also want to eat pigs, perceiving them to be an important bush resource. Cattle graziers control pigs to protect their water sources from becoming treacherous to cattle; in controlling pigs, graziers are enacting care toward cattle. Queensland Parks rangers control pigs in order to protect wetlands, waterways, biodiversity, and the flourishing of various native flora and fauna, and to preserve the visual amenity of the park so that tourists can encounter the kind of wilderness landscape they are expecting. Through instigating mass pig death, park rangers are doing the noninnocent and violent work that relations of care sometimes demand. Yet, these relations of care are complicated by the intensity with which Queensland Parks pursues and seeks to control pig hunters, indicating that only certain kinds of pig killing are deemed appropriate and only certain kinds of people are deemed appropriate to carry out pig control. Only *authorized* pig control is positioned as an act of care for land; *unauthorized* pig control, carried out by unknown pig hunters, is positioned as antisocial, destructive, and—ultimately—uncaring. In Queensland Parks, regimes of conservation and control overlap and coalesce, demonstrating how an institutional desire for regulation rubs up against the unruliness of the human-animal assemblages of the region.

PART III

Engulfing

Bushfire, Rinyirru National Park, 2020.

FIVE

Fire

On a pleasantly warm day in June 2018, I walked along the boundary fence between the jointly managed Lama Lama National Park and a cattle-grazing property named Tidewater Station, preparing to take part in my first planned burn. The air was still and calm. Around me, Lama Lama rangers positioned themselves, in pairs, at intervals along the fence line. Some rangers held drip torches, but most retrieved cigarette lighters or boxes of matches from their pockets. I was walking with Mabel, a Lama Lama ranger in her late thirties. Mabel bent down to snap a branch off a shrubby tree she called "bark tree." I had also heard people refer to this tree as "soap bush." It is a type of acacia, with long gray-green and silky leaves. Mabel handed me the branch and explained that I should use the branch to fan the fire if it seemed in danger of going out, or to extinguish any spot fires that jumped the road on the Tidewater side of the fence.

We spread out along the fence line and, before long, I could see telltale spirals of smoke from small fires nearby. Mabel handed me some matches and showed me how to light the dead grass near the fence. "Like this," she said, bundling the grass and lighting it at the base of the clump. She told me that this grass is called kerosene grass, presumably because it dries out, or cures, earlier in the season than other grasses and ignites easily. The fires that Mabel and I lit started trickling away from us, moving into the National Park. There was a satisfying crackling sound, and the smell of smoke. The fire crept through the grass, licking at the bases of trees but leaving much intact. Where the grass was still green, the fire was extinguished. Mabel kept an eye on me throughout the burning, reminding me intermittently to watch the fire. As we walked along, observing the progress of the fire and monitoring for spot fires, Mabel told me that she first learned to burn from her grandparents while they worked on cattle stations. Birds wheeled around in the smoke above us, waiting for the easy prey of panicked small animals fleeing the fire. We watched as the fire moved away. Once it had traveled around ten meters from the fence line, we packed up and returned to camp. Later in the evening, after several hours fishing on the banks of a small coastal inlet, I drove back to camp with several Lama Lama women. In the darkness, we could the see red glow of fires continuing to smolder. The women spoke quietly to each other, expressing their satisfaction that the burn had been a good one.

Fire is both a widely used land management tool and a threat to cared-for places, ecosystems, and livelihoods. Used for cultural burning to implement fire breaks and to reduce fuel loads, lighting fires is an act that is often explicitly framed as caring. In fact, across Australia burning is understood as one of the main ways that Aboriginal people care for Country. In this chapter, I explore how fire and care intersect and come apart. Among the people of Cape York, fire comes to work as a conduit for tensions among land managers, a way of aiming criticism across the land tenure and management divide. Land managers complain that their neighbors are applying too much fire, or too little; that their burning regimes are resulting in a reduction of biodiversity, or an increased risk of destructive wildfire; that firing the land is being done solely to take part in the carbon sequestration scheme. In both intentional burning and in the outbreak of wildfire, many people use fire to frame the actions of their neighbors as uncaring and to stake a claim about their own land management, their own enactment of care, and their own environmental values. However, as with other management issues, these criticisms are braided into broader critiques of land tenure changes and perceived government failings.

Burning to Care for Country

Burning is widely conceptualized by Aboriginal people in Cape York, and across all of Australia, as a way to clean up or look after the Country (Yibarbuk et al. 2001). This idea that burning is a tool for caring for Country is well established (Anderson 1985; Yibarbuk et al. 2001; Davis 2003; Head 1994; Ritchie 2009). Burning is often discussed as a key way for Aboriginal people to fulfill their obligations of custodianship to an area that has not been burned for a period of years and is often perceived as neglected (Anderson 1985, 81; Head 1994). Through burning, Aboriginal people enact care for the land and the ancestral spirits who inhabit the land. Historically, and today, fire is conceived of as a central way to socialize or humanize landscapes, to demonstrate care for the old people. As with other forms of land and environmental management carried out by Aboriginal people, care for the landscape, nonhumans, and the ancestral spirits who dwell in the landscape, is enacted practically—in the doing. Similar to weed control, the material practice of lighting fires, at the right time of year, is about showing respect and care toward ancestral spirits, reminding them that they remain significant in the lives of contemporary Aboriginal people.

Millennia of continuous burning regimes have shaped the distribution of species across Australia, resulting in a socialized landscape. In many parts of Australia, colonization violently ruptured and fragmented Aboriginal peoples' lives, disrupting fire regimes with flow-on effects for the makeup of ecosystems and functionality of landscapes. In regions where Aboriginal burning regimes have been significantly interrupted, fire-responsive and reliant vegetation has experienced a "dramatic decline" (Head 1994, 176; Bowman 1998). Such findings have added weight to a powerful argument that Aboriginal burning regimes have been an effective land management tool rather than being responsible for biodiversity loss (Langton 1998, 11).

Burning is also one of the more recognizable forms of Aboriginal land management to Western science. Evidence of historical and ongoing fire regimes has become a common way for Aboriginal groups to prove their current usage of land in land claims, and the right to burn has become a key concern when land rights or Native Title are recognized (Ritchie 2009, 48–49). Indeed, fire management tends to be the most straightforward way for Queensland Parks to conceptualize incorporating Aboriginal environmental knowledge into managing Parks. In Cape York, the responsibility to burn is shared by Aboriginal Traditional Owners, graziers, and Queensland Parks and, as noted by Davis (2003), each of these groups exercises their right to burn.

Controlled burns are used to achieve a variety of outcomes: as a cattle management tool to encourage the growth of pasture, in order to reduce the risk of damaging wildfires, and to maintain fire-adapted ecosystems. For Aboriginal people in Cape York, burning has an additional use in asserting cultural knowledge around land management and operating as a cultural marker. What people are referring to when they speak about burning in a traditional or cultural way is somewhat ambiguous and situational. For some Aboriginal people, like a Lama Lama woman named Irene, burning with traditional methods means having a senior person present while burning is carried out. While burning "traditionally" mostly seems to mean burning from the ground, some Aboriginal people in Cape York also see aerial burning in which a Traditional Owner is in the helicopter as equally traditional. In this sense, traditional burning can be used to refer to any kind of fire management that has input and involvement with Traditional Owners. Burning, then, is as much a cultural activity as it is an important aspect of land management. In suggesting that fire management is "traditional" when Traditional Owners are involved, Irene is asserting that burning is a cultural activity that, when done in certain ways, is something distinctly Aboriginal.

Many Cape York Aboriginal people refer to the way that the "old people" would burn Country, highlighting both the economic purposes of burning and the long history of fire management in Aboriginal communities in Cape York. As Peter, senior ranger for the Rinyirru Aboriginal Corporation, told me,

> You know, our old people would burn at that first wet burn. Try to burn most before the drought come in. So you haven't got much fuel there, it keeps all the animals in that one area, the grass, birds, bees. Different area got different flowers for good sugarbag honey and English bee honey. So they knew what time to burn that Country. And the wild rice, to bring it back good. . . . They knew when to burn all those Country.

Reflecting similar sentiments, John Bradley has suggested that for the Yanyuwa people in the Gulf of Carpentaria that he has worked with, burning is a deeply cultural activity that allows people to "demonstrate[e] a continuity with the people who have died, their ancestors" (1995, 28). Yet, in Cape York, the transmission of knowledge has not been a linear trajectory from older, experienced Aboriginal people to younger.

While several Aboriginal rangers spoke to me about learning fire management skills from their parents and grandparents, every ranger nowadays must complete a formal nationally recognized certificate in fire safety and management. Many of these courses are facilitated by CYNRM. I attended the final day of one of these training courses with several of the Rinyirru Aboriginal Corporation rangers in mid-2018. On this day, people from an assortment of local Aboriginal ranger groups were gathered at a community building in the town of Laura. For the first half of the day, the CYNRM trainer discussed weather conditions, fire behavior, different types of burning, and safety precautions. Throughout, he acknowledged that what he was imparting was likely "old news" to the Aboriginal rangers present, reflecting his awareness of the experience that many of these rangers already had in lighting and fighting fires, and deferring to their status as cultural fire managers. In the afternoon, the group undertook a controlled burn in a vacant lot in the town, each of us positioned at points and allocated particular tasks, from starting the fire using drip torches, to operating mop-up units (portable water tanks), to observing the edges of the fire to ensure that it did not spread beyond the intended area.

For those Aboriginal rangers who attribute their fire-management knowledge to their parents or grandparents, it is important to note that these parents and grandparents worked on cattle stations. It is here that they engaged in fire management, and, in many cases, here that they learned how to manage fire from older Aboriginal stock workers. The intercultural context of the grazing industry and mutually beneficial outcomes of early dry season burns, colloquially called cool burns, for Aboriginal people and graziers means that the boundaries between Aboriginal fire management and grazing fire management are murky.

Cool burning the melaleuca scrub, Lama Lama National Park, 2018.

Cool burning, Lama Lama National Park, 2018.

Burning on Cattle Stations

The way that graziers burn has similarities with the way that Aboriginal Traditional Owners burn. Graziers told me about lighting fires by riding along slowly on a quad bike, holding a lit drip torch in the grass, and leaving a trickle of fire in their wake. Instead of leaving fires to burn mostly unmonitored, graziers tend to keep track of fires using an online resource called Northern Australia Fire Information (NAFI) and by checking the horizon for visible smoke. This is a tool that CYNRM, Queensland Parks, and Aboriginal ranger groups also use.

In general, graziers have the view that burning from the ground is the most appropriate way to burn. While aerial burning can be fast and effective for burning large swathes of country, it is a method that relies on proper firebreaks having already been put in place. One grazier, Martha, pointed out that it is more difficult to keep track of which sections you have burned when you burn from the sky. The prevailing logic with burning is that each section of country should be burned every three or so years. Aerial burning means that it can be more difficult to achieve the kind of mosaic burns, where discrete patches of land are burned and nearby patches are left to rest, that are desired. Another grazier, Bev, echoed Martha's sentiments. Bev told me that she thinks that the way Queensland Parks burns—frequently from helicopters—is wrong, as it often results in fires meeting each other and leaving nowhere for the animals and insects who live in those areas to move to safety. Bev and her husband, Alan, prefer to burn from the ground, believing that this style of burning provides animals a chance to get away.

Several graziers spoke about mosaic burns as being important to allow the country to rest in between times of being burned. This kind of mosaic burning that graziers carry out has substantial overlap with the kind of burning that Aboriginal Traditional Owners do. The key difference is the involvement of Aboriginal Traditional Owners, although up until a handful of decades ago, grazier burning consistently did involve Aboriginal people, as they were the bulk of the workforce on cattle stations. An employee of CYNRM pointed out this convergence in burning regimes, saying that the way graziers burn is "very similar to what we think was happening 200 years ago. You know, like talking to people and stuff, you sort of see . . . talking to the old people, Aboriginal people were always walking around with a firestick. So, they were burning." When asked, graziers invariably spoke of learning how to use and manage fire from their parents. However, it is well documented that pastoralists in parts of the Northern Territory incorporated Aboriginal burning regimes into their own land management practices (Ritchie 2009, 46). It seems likely, then, that even if graziers learned about fire from their parents, their parents' knowledge originated with Aboriginal land managers in the past.

While graziers generally concede that cool season burns are necessary to protect pasture and infrastructure, to encourage the growth of fire-adaptive plants and as a cattle-management tool, land managers have a range of perspectives on the potentially detrimental impact of exclusively lighting cool fires. One grazier, Bill, told me about the transformation of the landscape on the station where he was raised once it became a national park and the fire regime shifted. He recalled a visit to the area to do some contract mustering, having not seen the station for twenty-seven years. He recounted his impressions to me:

> I drove into [the station] and there's a ridge, probably about two kilometers[1] from [the] house, there's a big ridge there. And that used to be our paddock where our stallion and mares, our breed mares were, for breeding workhorses. . . . And you could ride up on that big sand ridge there and look, and there's a horse three or four hundred yards, and you could see it. But that ridge today, it's—it's got a lot of white currant out there which usually grows in at the river, and that's gone right out there. Even when I was there that day, it was nearly dark and there was a little cold fire trickling through this stuff, and I'm thinking, well that's going to be more and more. That open savanna country, as they call, well that's going to be no more. And . . . yeah, that's . . . I don't know. I just think this whole fire thing is wrong. Totally wrong.

Like other graziers, Bill implements cool burns. However, he asserts that cool burns alone do not constitute an effective fire regime. This is because these fires are not hot enough to properly "clean up" the country. Moreover, cool burns are understood by land managers to result in thickening of the country by encouraging the growth of melaleuca scrub, referred to as woody suckers. Melaleuca encroachment is considered undesirable because it transforms landscapes from open savanna grasslands to dense scrubby forest. This is an issue from both a conservation standpoint and grazing point of view, as it affects groundcover, habitat for small animals, grasses, and makes it more difficult to run cattle.

Most people share this perspective, even as they engage in lighting fires of this nature and sometimes, through the Australian government's carbon sequestration scheme, profiting from them. Only one grazier revealed to me quietly that he did not believe cool burns were as bad as people claimed. He told me that he is not so sure about the kind of burning they carry out on his own station. He said that everyone is always talking about the problems with cool fires, but he has seen cool fires burning along "slow and steady, doing good work." With a chuckle, he admitted that the "experts"—in this context, his wife—do not agree with him. This particular grazier had lived on his station his entire life and told me that he has seen woody thickening happen in areas that have never been burned and never had cattle. He suspects that woody thickening is perhaps just a natural tendency of the landscape and less connected to cool burns than most would argue.

Despite the reservations that some graziers hold about the impacts of implementing cool burns, it is a form of land management undertaken across the region. As I have described, the type of burning undertaken on pastoral leases bears similarities with the types of burning carried out by Aboriginal Traditional Owners. I contend that it is unlikely that this has occurred as a linear transmitting of Aboriginal knowledge to settler-descended graziers. Aboriginal burning regimes, while continuing from precolonization to the present day, have undergone shifts, adaptations, and changes in purposes and practices both before and after non-Aboriginal settlers arrived. This is nowadays compounded by the introduction of aerial burning and the formalizing of fire-related training.

I conceive of the historical grazing industry as a site of productive yet uncomfortable relating. The collaborations between Aboriginal people and white graziers resulted in graziers gaining important knowledge in managing their land, while allowing Aboriginal Traditional Owners to maintain a physical relationship to their ancestral lands—albeit in a highly restricted fashion and reliant on their continued employment for little or no wages. Through this space of intercultural interaction, graziers gained knowledge that enabled them to care for land in ways that have some commonality with Aboriginal forms of land management, and Aboriginal people were able to continue to work on, live on, and fire their areas of traditional connection. The environmental knowledge that underpins the contemporary burning practices of Aboriginal Traditional Owners, graziers, and even Queensland Parks rangers has emerged out of a process of coproduction and continues to develop in new ways as a result of new entanglements.

As such, it is problematic to conceive of "Aboriginal burning" and "grazier burning" as two separate domains. Everyone implements cool burns, whether to reduce fuel loads, participate in carbon sequestration, or encourage the fresh growth of "green pick" and draw cattle to particular areas. Through ongoing processes of interaction between people, plant and animal species, technologies, and government-led initiatives, burning knowledge continues to be coproduced over time. The formal fire training workshops reflect this confluence of different genealogies of knowledge, drawing on the rhetoric of cultural burning alongside the language of hazard reduction. Though some of the intended uses of fire differ from land-tenure type to land-tenure type, accepted knowledge and wisdom about how fire operates and should be managed in Cape York is held in common.

Burning in National Parks

In Rinyirru National Park, burning occurs a little later in the year. I was in the park in July when some aerial burning was carried out, involving rangers Ray, Sammy, and Mitch. Ray is the ranger in charge for the park, Sammy is a conservation officer for Queensland Parks whose role involves a lot of fire management, and Mitch is a helicopter pilot. Sometimes Queensland Parks involve Aboriginal Traditional Owners in their aerial burning, but on this occasion no Traditional Owner was present. Sammy explained to me how aerial burning works. He told me that in the helicopter, the bombardier sits beside a machine that injects small balls called Dragon Eggs with a flammable gas and drops the balls into the scrub. The Dragon Eggs take about fifteen seconds to ignite and burn for around twenty seconds, just long enough to start a small fire. Normally, the machine is set to drop the gas-filled balls every twenty or fifty meters as the helicopter flies along.

Ray explained to me that he thinks of cool burns as protection burns and is concerned mostly about establishing adequate firebreaks along the boundary fences in order to protect neighboring properties and Queensland Parks infrastructure in the event of a wildfire spreading through the park. As part of Queensland Parks' fire management, a fire plan is created each year and approved by Queensland Parks' management structure and the board of Rinyirru Aboriginal Corporation. Before any fires are lit, Queensland Parks' managers are required to give an "all clear," and neighbors must be notified. Throughout this process, though, the ranger in charge remains the decision maker, a situation that sometimes draws the ire of individual Aboriginal Traditional Owners.

On this day, Ray announced happily that the neighboring grazier has given him the go-ahead to "burn the shit out of it" by helicopter. Ray told me that while July is generally the latest they are able to burn in the dry season due to the risk of fires later in the year spreading uncontrollably, this year they would need to carry out additional burns later on, as everything was still a little too green to burn effectively, owing to a later than usual wet season in the preceding months. After the fires were lit, Ray and Sammy sat monitoring a laptop screen. They had loaded the same NAFI website that graziers use to keep track of fires. NAFI displays information about the location, intensity, and duration of fires, and is color coded to show where fires have burned in previous weeks, months, and years. Sammy monitored the website to see whether their fires were showing up and still burning. He explained that he was concerned that the vegetation was not cured enough for an effective burn.

Despite these efforts to burn, Ray instigated a further attempt to burn in August. This time, Queensland Parks rangers were to burn from the ground. August is widely considered to be too late to burn in Cape York, but Ray explained that he was eager to put a firebreak in place to protect the main ranger base from wildfires. At the fence line, some of the rangers walked into the scrub carrying drip torches and began lighting up the grass. Others were in vehicles equipped with firehoses and large water tanks, colloquially called mop-up units. These vehicles were stationed at intervals along the fence line, ready to intervene if the wind changed direction. From my vantage point in a ranger vehicle, I watched as the fire began. At first, tendrils of smoke rose above the trees. Before long, enormous clouds of smoke enveloped me. Two rangers were stationed with a mop-up unit beside the fence, only a few hundred meters away. At points, the smoke was so thick that they were difficult to make out. Once the fire had spread through much of the western side of the patch of scrub, Ray's voice crackled through the two-way radio. "That's good, we'll pack up," he said. We regrouped on the main road that ran on one side of the burned section of land and spent the next few hours monitoring the fire. While some of the Aboriginal rangers present were vocally critical of the decision to burn so late in the season, this hazard reduction burn was ultimately conceded as necessary when intense wildfires swept through the park in December.

Queensland Parks fire-management regimes intersect with genealogies of fire knowledge that constitute burning regimes elsewhere in Cape York. For instance, Rinyirru National Park's fire regime is an adaptation of the types of burning that their grazier neighbors employ. I spoke with the former ranger in charge for Rinyirru National Park and current (at the time of my field research) CYNRM Landcare officer, Michael. He explained that when he first came to work at the park, he based his burning regime on what was being implemented across the boundary fence at a neighboring cattle station. As well as learning from the neighboring graziers, Michael sought the advice of several Aboriginal Traditional Owners who were senior knowledge holders. These men had worked alongside the aforementioned neighboring grazier in the mustering camp, passing on knowledge to him that he still holds today.

As should be clear from my descriptions of fire management, while Aboriginal people, graziers, and Queensland Parks diverge in aspects of their burning practices, there are also significant overlaps and mutual influences. Indeed, Ockwell and Rydin have argued that these groups in Cape York constitute something of a "pro-burning coalition" with "inter-connected storylines" (2006, 394). In some sense, the various kinds of burning regimes in Cape York stem from a common origin: precolonization Aboriginal burning regimes. However, forms of fire management have been adapted to suit different purposes and priorities, incorporating aspects of Aboriginal burning knowledge, grazing burning practices, and Western scientific approaches toward the use of fire in landscape management.

The uptake of burning regimes in national parks demonstrates how the project of joint management has, to some extent, functioned to incorporate Aboriginal fire management—a form of what is often referred to as traditional ecological knowledge (Berkes 1993; Nadasdy 1999).

A ranger oversees a hazard reduction burn, Rinyirru National Park, 2018.

A ranger vehicle along the fire line, Rinyirru National Park, 2018.

Elsewhere in the world, the reintroduction of indigenous fire regimes remains contentious.[2] In northern Australia, however, the incorporation of fire regimes into Queensland Parks' management plans for the Cape York region is not particularly contentious today, even if tensions emerge in how fire regimes are carried out. This is the result of several factors: decades of social science and ecological research supporting burning regimes in northern Australia (Bradley 1995; Bowman 1998; Yibarbuk et al. 2001; Perry et al. 2018); the continuity of fire regimes from precolonization to the present day (albeit with some disruption) resulting in a landscape that still responds positively to fire (Head 1994; Langton 1998; Davis 2003); the use of similar fire regimes on neighboring properties; the introduction of fire regimes in the park before the era of joint management; and the formal structure of joint management itself, which, however imperfectly, seeks to take Indigenous environmental knowledge claims seriously (Reardon-Smith 2024). In Cape York, joint management (though still in its "teething" phase) works against the fortress conservation model, as it incorporates the perspectives, practices, knowledges, and values (albeit in a constrained fashion) of Indigenous peoples who are Traditional Owners for protected areas. While the ontological "building blocks" of management (Howitt and Suchet-Pearson 2006) may remain consistent with pre-joint-management conservation, the use of fire regimes based on precolonization burning represents perhaps the most successful and visible incorporation of Aboriginal knowledges into park management.

Both the grazing industry and the project of joint management are sites of collaboration that have been simultaneously enabling and constraining for Aboriginal people. While Queensland Parks burning was initially modeled on the already interculturally mediated fire management undertaken on nearby pastoral stations, it is now legislatively mandated to involve the engagement of Aboriginal Traditional Owners. In turn, Queensland Parks now provides resources and training for Aboriginal rangers to carry out burning. The uptake of burning regimes and delivery of fire training by Queensland Parks reflects this deeply entangled intercultural dynamic. Yet, despite the involvement of Aboriginal Traditional Owners, it is Queensland Parks' management staff who formulate a fire management plan and the ranger in charge who makes the decision of when to implement burning regimes. Such a situation results in some Aboriginal Traditional Owners articulating a sense that their knowledge and perspectives have been sidelined by Queensland Parks management. Senior ranger for the Rinyirru Aboriginal Corporation, Peter, expressed such an opinion to me. He said that Parks management is happy to be "just jumping in and just burn. Burn what you can. It's the fuel, to save visitors at campsites when the wildfire [comes] and all that kind of stuff." To Peter, burning for hazard reduction alone, and burning too late in the season when the fires do not encourage the growth of different plant species, is a departure from Aboriginal fire knowledge.

Fire-management and burning regimes are not simply a local concern with regionalized impacts in Cape York. The introduction of the carbon sequestration program has drawn Aboriginal Traditional Owners and graziers into engagement with the international carbon trading market, although the carbon sequestration program tends to be understood in a deeply localized way in Cape York. With carbon credits considered to be both lucrative and problematic by many land managers, further tensions emerge.

Carbon Sequestration

This sharing and co-creation of fire knowledge between Aboriginal Traditional Owners, graziers, and Queensland Parks rangers is complicated by the existence of the carbon sequestration program. The carbon sequestration program was first established in Arnhem Land in 2005, and since then has been implemented across northern Australia (Russell-Smith et al. 2013). Engagement in the carbon credits scheme requires burning to occur each year between the first of January and the thirty-first of July. Any fires that occur after this time frame, whether deliberate or accidental, threaten the payment that the land manager will receive for burning. The payments that land managers receive do not come directly from the government but through a third party with whom the land manager has a contract.

The logic behind the carbon credits scheme is based on reducing carbon emissions. Because of the higher fuel load later in the year, as more grasses cure and other vegetation dries out, late-season fires burn hotter, potentially burning larger trees and spreading into the canopy, and release about twice as much carbon dioxide into the atmosphere as an early-season cool burn. Late-season fires also tend to be significantly larger. The carbon credits program is intended to encourage early-season burns, which establish firebreaks and reduce the fuel load, thus avoiding out-of-control wildfires later in the season. The financial rewards for carbon credits can be substantial, leading to criticism from various Cape York locals that certain people are damaging the landscape with excessive fire for profit. The carbon credits scheme in Cape York, but even more broadly across northern Australia, can be read as an attempt to remunerate Aboriginal people for caring for Country and employing traditional land management practices. For Cape York locals, though, this interpretation was rarely considered—probably because settler-descended graziers are equally entitled to engage in the carbon credits program.

There are varying opinions among Cape York locals about the value and efficacy of the carbon credits program. It is widely conceded that the carbon credits scheme does deliver some benefits in that it encourages and creates capacity for regular burning that may not have been happening previously. During a discussion between grazing couple Bill and Diane, Diane pointed out that since the introduction of carbon credits, hot, out-of-control bushfires have become less frequent. She recalled the regularity of such wildfires when she and Bill first acquired their station, some thirty years earlier, saying:

> We would've burnt when the storms were coming. Like, when we first came here we got burnt out, black, because we didn't have any roads, we didn't have anything. We got burnt black every second year. Just the whole place just burnt, you know, before we did that. But now it seems to be a bit more, not so much . . . I think that's what they were trying to stop was the out-of-control late season fires.

Similarly, the Landcare officer for CYNRM, Michael, pointed to the positive impact on the landscape that carbon credits have had. He told me that the impact is perhaps more notable on the western side of Cape York, which historically has often burned later than the east. Michael explained: "that's where the prevailing winds go, and if a fire starts anywhere on the Peninsula Development Road late in the year it used to keep going until the west coast." Michael also pointed out that substantial tracts of land in that region of the Cape are Aboriginal land.

Importantly, the carbon credits scheme has provided both the financial incentive and the capacity for Traditional Owners in that region to burn their Country early enough to create firebreaks to reduce the spread of wildfires.

Carbon credits is often referred to by graziers as "fairy dust money" because it is considered to be "money for nothing." When I initially heard this term, I assumed the grazier in question was denigrating the carbon credits scheme as another pointless government-directed initiative, dreamed up down south and awkwardly applied to Cape York. In conversation with Michael, though, I understood that referring to carbon credits as "fairy dust money" is related more to graziers' bemusement that they could be financially rewarded for something they have always done, rather than a criticism of the program. The majority of graziers I worked with in Cape York engaged in the carbon credits program. Those who did not were unable to do so because of complex limitations around their particular lease agreements rather than because of an ideological position. This gestures toward the radical economic contingency of the region (Neale 2017; Reardon-Smith 2023a). Even those graziers who do engage with the program are critical of what the carbon credits program is actually achieving.

There has been compelling scholarship on how carbon sequestration projects, particularly in the Global South, have transformed local lives and livelihoods (Blok 2011; Dalsgaard 2013; Yocum 2016; Valderama 2020). Much of this work details how these projects, precipitated by multinational actors or NGOs, seek to change local practices with a goal to both addressing climate change and contributing to local development, with varying levels of success on both counts (Yocum 2016). As Blok (2011) points out, these kinds of projects then shift the responsibility of carbon emissions and mitigation onto some of the people who have contributed the least to anthropogenic climate change. The situation in Cape York is both different and the same. In some ways, carbon sequestration in Cape York (and across northern Australia) has emerged as a novel way to pay people for the labor that they already do—or at least have been aspiring to do, which is the case for some Aboriginal traditional owner groups who previously lacked the resources to implement wide-scale early-season burning. Indeed, as noted by Timothy Neale (2022, 11), Indigenous carbon credits are understood by carbon trading companies as a kind of "premium" product due "their ostensible alignment of Indigenous empowerment and climate change mitigation and, thereby, their virtuousness." In this sense, carbon sequestration is less a transformation of local practices and more a facilitation or recognition of the work that, in particular, Aboriginal Traditional Owners (but also cattle graziers) have already been doing. Unlike the carbon sequestration projects in the Global South, in northern Australia these schemes do not seek to transform local relations to land or livelihoods as such. However, in a smaller but nonetheless significant way, carbon credits do transform local burning practices to an extent, and it is these changes that are particularly contentious among land managers.

There was a general sense among graziers—and some Queensland Parks rangers—that carbon credits encourage the wrong type of fire management by providing payment for cool burns, which may lead to melaleuca encroachment, otherwise called woody thickening. Carbon credits can also provide incentives for land managers to burn substantial swathes of their country each year, instead of engaging in mosaic burns that allow each section of Country to rest in between burns. Many graziers neighbor parcels of Aboriginal freehold land that do not operate as active cattle stations. According to these graziers, their Aboriginal Land Trust neighbors get much of their income from the carbon credits scheme and tend to overburn, with ramifications for woody thickening.

The same critique was put forward by several Queensland Parks rangers at Rinyirru National Park. They told me that a neighboring Aboriginal Land Trust had burned far too much land over a short period of time, rushing to get their burning completed before the carbon credits cut-off date at the end of July. The Queensland Parks rangers I spoke to characterized such burning as unnecessary and irresponsible, motivated purely by financial incentives rather than a desire to care for or clean up the Country. Burning motivated by the carbon sequestration scheme is framed by detractors as harmful, and as uncaring.

The narrative that carbon credits allow Traditional Owners to be financially rewarded for caring for Country is complicated by the reality on the ground. Not only does the carbon credits scheme encourage a different and less beneficial type of fire regime than might have existed previously, but the burning itself is frequently carried out by a Landcare or CYNRM employee in a helicopter. Thus, in some ways the carbon credits scheme can be understood as precipitating a new type of burning regime rather than paying people for the work they already do, and this is not only the case for Aboriginal Traditional Owners. Graziers, too, admit to such a tendency. Such a sentiment was summed up succinctly by one grazier who, when asked if he believed carbon credits has changed the way people burn, replied, "Oh shit, big time! I even know people who know better, will say, oh well I need that money."

Despite the carbon sequestration program being premised on mitigating carbon emissions because of a changing climate, the concept of climate change rarely came into conversations about fire. In Cape York, controlled burning and wildfires have always been a reality for people living on the land. The impacts of a longer, less predictable and more catastrophic fire season are not being felt in Cape York the way that they are being felt in the southern part of the Australian continent. For many graziers in Cape York, who tend to characterize shifts in weather patterns as "natural cycles" (Connor and Higginbotham 2013; Connor 2016), the concept of climate change represents an ideology or belief system, rather than a reality. Thus, the purpose of the carbon sequestration program—ostensibly to reduce carbon emissions—is obscured or ignored by many of the people who engage in the program. Instead, the program is discussed by graziers and Queensland Parks rangers, in particular, to criticize some Aboriginal groups' burning regimes, which are perceived as being detrimental to the Country.

However, as is evident in the way that graziers simultaneously critique and take part in carbon credits, it is difficult for financially marginal people to reject the offer of these payments. The same is evidently true of Aboriginal land trusts that are criticized by their neighbors for burning too much land for profit. The power that financial reward holds encourages land managers to engage in the program, even as they continue to harbor concerns. The carbon credits scheme provides a route to maintaining fire regimes as it remunerates land managers for their work, but it is constraining in that it encourages only one kind of burning. Furthermore, in this reification of what constitutes an appropriate burning regime for northern Australia, the various inconvenient impacts of exclusively implementing cool burns are glossed over.

Engagement in the carbon credits system is ubiquitous among eligible Cape York land managers, because, as grazier Bill pointed out, people "need that money." Enacting proper care for the land, here, rubs against the financial precarity that many people in Cape York experience. Despite the claims of Cape York land managers that carbon credits is fairy dust money, or, money for nothing, at least some people would contend that carbon credits has transformed the way that people burn and has shifted people's priorities for burning. Instead of burning to cultivate a particular kind of preferred landscape, some graziers and Aboriginal groups are compelled to burn more than they normally would to ensure that they receive payment. Using fire to maintain particular landscapes for particular purposes, though, remains a concern—particularly for graziers who rely on pasture for their economic well-being. Many graziers implement storm burning to lessen some of the effects of cool fires despite the risk this poses to their receiving carbon credits. This is a type of burning that some people consider to be vital in maintaining the landscape, providing insights into what constitutes different kinds of preferred landscapes for different people.

Storm Burning

Toward the end of the year, I drove through a forested section of Lama Lama National Park with graziers Alan and Bev who lease Tidewater Station, situated between Rinyirru National Park and Lama Lama National Park. It had been a while since they had driven on this road, as bogginess had rendered it impassable until late in the dry season. As we drove through the forest, Alan and Bev observed that "the country ha[d] thickened up a bit" since this section of land was excised from their station some ten years prior. Alan and Bev explained to me that the woody thickening here would have been much worse had they not been able to continue implementing storm burns for some years after the land transfer.

In order to counter the issue of woody thickening or melaleuca encroachment, which most people understand as related to cool burns, most land managers also engage in another type of controlled burning called storm burning. Storm burns can occur only within a tight temporal window, after the first rain has come, but before the wet season sets in properly. Some years there are only a handful of days in which it is appropriate to storm burn. Land managers understand storm burns to be essential to counteract the woody thickening that cool burns encourage, as these fires are hot enough to burn "suckers" (melaleuca saplings) and are effective for cleaning up the country and maintaining open grasslands. Pam, an experienced grazier, explained to me why she does storm burning:

> You need a hot fire to control the regrowth or thickening—whatever you want to call it . . . To reduce the thickening. Our problem is we've got too many trees. We shouldn't have this many trees. It should be a lot open-er. And we've got to keep it open. The only way you can keep it open is to have really hot fires in the dry weather. And not in the proper dry, just before the storms. It's got to be big humidity.

Storm burning is an attempt to counteract some of the consequences of the widespread implementation of cool burns. While many land managers engage in both types of burning, several graziers expressed to me their belief in the primacy of a prehuman "natural" fire regime. Grazier Bill, who actively implements a burning regime involving both cool burns to establish firebreaks and storm burns to maintain open savanna landscapes, discussed his belief that Aboriginal burning regimes, grazier burning regimes and, now, the carbon sequestration program have disturbed a "natural" and preferable type of burning. Early on in our interactions, Bill voiced his belief to me that before either European settlers or Aboriginal people were in Australia, the only way the country would burn was from lightning strikes—that is, late in the dry season, immediately before the wet, when dry electrical storms sweep across the region as the monsoon builds. Over the course of many conversations, he elaborated on his perspective:

> I just think we're going dead opposite to nature. Nature, before anybody interfered as far as I'm concerned, was hot fires, storm time, when lightning struck. You know, that was before whitefella or blackfella, anyone interfered. And that's my argument. Why are we going dead opposite to nature with the burning? Why are we burning now? Because no way in the world [would] nature have burnt this time in the year. It couldn't. It only burnt come November . . . October, November, December, when lighting come around. And I just think this is really wrong, this early burning. That's what's thickening the country up.

These ideas of a "natural" fire regime are shared by a variety of graziers. Another grazier, Martha, told me that she believes "nature" intends for fires to occur in the late season and burn hot, and this is what the country relies on to maintain open savanna landscapes. "I reckon nature's been around a lot longer than any people," Martha reasoned to me. Other graziers shared this belief in a natural fire regime and suggested that Aboriginal people likely "messed up" the natural way of things with their burning, whereas hot, lightning-strike fires would have kept the country open.

Graziers are drawing a distinction between "nature" and "culture," which goes back to the key ontological and epistemological assumption in Western thought that nature and culture are separate. Today, it is more or less accepted in the social sciences that both nature and culture and the relation between them are constructed in particular ways to serve particular ends (Descola and Pálsson 1996; Descola 2013; Escobar 2008; Pálsson 1996). Descola and Pálsson (1996) argue that the ontological category of nature, as separate from culture, is unique to Western societies. They argue that despite notions of "wildness" existing in some form in many societies, what is significant to a Western ontology and epistemology is the reliance on a binary framework: nature in opposition to culture. These ontological assumptions underpinned the preservationist model that national parks are based on.

As I have already noted, many scholars have criticized the assumptions that such a preservationist model rests on, pointing out that the landscapes that governments may wish to protect from people are already socialized and culturalized landscapes (Cronon 1996; Head 2000; Balée 2013). That is, they have already been altered, transformed, exploited, and nurtured by people over long periods of time, resulting in contemporary landscapes that may appear natural but are in fact the result of human effort, labor, and care. Marcia Langton writes that,

> The Aboriginal objections to the term "wilderness" do not, of course, constitute an objection to the protection of natural values, but rather a demand for recognition of the cultural content of biophysical landscapes and the extent of the interdependence of cultural and natural values, at least in those landscapes where there has been almost uninterrupted Aboriginal management for millennia. (1998, 10)

This is very much the case for Cape York, particularly in relation to the importance of continuing fire regimes in managing vegetation and landscapes. While graziers are not disputing that Aboriginal people have socialized the landscapes of Cape York through burning regimes, they are suggesting that such burning regimes are "going dead opposite to nature." Graziers position both historical and contemporary fire management to be a human intervention that has caused unwanted changes to the landscape. Yet, graziers continue to fire the landscape.

The notion that storm burning is necessary hinges on two related ideas underpinned by particular sets of knowledge and ontologies: that Cape York ought to be composed of open savanna grasslands, and that the only prehuman (and thus natural) fire would have occurred from lightning strikes during storms. What mostly goes unsaid is that open savanna grasslands are preferable for grazing cattle. The fact that graziers understand a cattle-appropriate landscape to be a more natural landscape is related to how graziers frame the landscape of Cape York and their role in it in general. The notion that cattle and the landscape in Cape York are a natural fit emerges from the value that graziers see in laboring on and with the land and the centrality of cattle to their ability to live and work in Cape York.

Graziers' sense of belonging to the region is deeply entwined with their physical laboring on the land, and this labor is mediated by their relationship to cattle. For graziers, their way of life—as graziers, as people from the bush—is related to how they understand their cultural place in the world. Graziers' existence in the world is entwined with cattle on multiple levels. Of particular importance is how graziers come to experience landscapes as mediated by cattle, in a similar vein to how both Dominy (2001) and Gray (1999) describe sheep as shaping pastoralists' relationships to land. Graziers tend to talk about landscapes, weather conditions, and vegetation in terms of how they may affect cattle. Similarly, what they perceive to be a "good" landscape is one that is conducive to grazing cattle—a landscape free of melaleuca encroachment. Graziers' relationships to cattle and cattle-appropriate landscapes are not purely economic but are instead about a particular valued way of being in the world (see Reardon-Smith 2021). By suggesting that such a preferable landscape is the result of a natural fire regime, graziers dismiss the changes caused to ecosystems by extensive and long-term burning regimes. Yet, despite this belief in a natural fire regime, many graziers both engage in burning and understand the disruption to natural fire as achieved by human activity. Only the sole grazier in this study who disputed the dominant narrative around cool burns and woody thickening seems to perceive human fire regimes as somewhat irrelevant.

Storm burns were not discussed with me in detail by either Aboriginal Traditional Owners or Queensland Parks rangers. While some Queensland Parks rangers who were concerned about the impacts of melaleuca encroachment did position storm burning as necessary, though tricky to implement because of the risk-averse nature of Queensland Parks management, in general storm burning seemed to be a preoccupation for graziers. This is because even though Queensland Parks, Aboriginal, and grazier fire regimes look similar and result in similar outcomes, the ontological basis and purpose for each fire regime differs. Fire regimes on graziers' properties are ultimately intended to preserve pasture. They implement cool burns and mosaic burns to encourage the growth of grass and to establish firebreaks that will save their pastures from being destroyed in out-of-control hot fires late in the season.

Because these cool burns are understood to encourage melaleuca thickening, graziers also implement storm burns to try and allow the landscape to remain open. While they position this as an attempt to maintain the natural savanna landscape of Cape York, this is evidently preferable to woody thickening because it allows for more pasture to grow and for graziers to more easily and economically grow cattle. Graziers' practices are related to what they understand a natural fire regime to be, albeit in a controlled fashion in order to produce a particular outcome. In their storm burning, they seek to mimic what they believe was happening "before whitefella or blackfella, anyone, interfered." Thus, there is a tension here between the hierarchical relationship they construct between natural and cultural fire. However, in practicing storm burning, the binary between "natural" and "cultural" fire—and between nature and culture in general—that graziers have constructed becomes unstuck.

Wildfire

On a quiet afternoon at Silver Plains Station, I was spending time with the Lama Lama rangers. Most of the rangers were gathered around the homestead, casually chatting, when senior ranger Patricia tore into the yard in her Toyota. Jumping out of the car, she quickly called out to gather the rangers. She announced that there was a fire at Marina Plains, a section of Aboriginal freehold owned by Lama Lama Land Trust but leased to graziers Alan and Bev for their cattle. Patricia said that any of the rangers gathered who had obtained their certificate in fire training were to accompany her to assist Queensland Parks rangers in containing the blaze. There was a flurry of activity as the handful of qualified rangers gathered the equipment they needed before departing with Patricia.

Because the distance between Marina Plains and Silver Plains is quite significant, they stayed in the motel-style accommodation at Musgrave Roadhouse, an hour or so inland from Marina Plains. When they returned, I spoke with Mabel, one of the rangers. She told me that they had had an exhausting few days fighting the fire. She said that along with the Lama Lama rangers, some Queensland Parks rangers from Rinyirru National Park were present, as well as Alan and his son, Brad, from Tidewater. Mabel added that Alan was seemingly only concerned with protecting his own paddocks and pasture, caring more about conserving grasses than anything else.

Later, when I visited Tidewater to see Alan and Bev, they recounted the same event. Alan said that he had spotted smoke south of Tidewater and had gone to investigate, realizing once he was closer that the Rinyirru National Park side of the boundary near Marina Plains was alight. He had phoned Queensland Parks to inform them about the fire. The ranger in charge of Rinyirru National Park had replied that he knew about the fire and was keeping track of it using the NAFI website. When Alan informed the ranger in charge that the fire was actually on the park side of the boundary, Queensland Parks rangers, Lama Lama rangers, and Alan and his family converged on Marina Plains to fight the fire. Alan told me that he had been disappointed with the Queensland Parks rangers' attitudes throughout the event, saying that these rangers had been solely focused on protecting Queensland Parks infrastructure and had shown little concern for the paddocks that were at risk of being burned out. Alan told me that the rangers had repeatedly said that the park was their priority. Alan said that, from his perspective, fighting fires ought to be a joint effort in which everyone seeks to protect as much as they can from the fire—fighting the fire alongside each other as neighbors, with any other politics suspended. Of course, from what Mabel said, Alan had his own priorities in terms of what he wanted to protect.

Another grazing couple, Martha and Gerald, told me about their experience fighting fires toward the end of the dry season. Martha explained to me that from around October onward, they, along with everyone in the region, are pretty much occupied with monitoring and fighting fires. Martha and Gerald's station, Ironwood, is situated on a plateau that is part of the Great Dividing Range. Their station is particularly vulnerable to fire, although, as Martha pointed out, everyone in Cape York is vulnerable to fire. At this time, Martha was the fire warden for her area. She told me that she would travel to a particular ridge on her station, from which she had a good outlook, to keep an eye on fires. There is a general consensus that when fires are burning, everyone who is local ought to pitch in to help fight fires. Martha and Gerald relayed a disappointing recent experience during which they went to a neighboring property to help fight a blaze. Despite the neighboring Aboriginal Corporation and Queensland Parks supposedly being aware of the fire, none of these groups provided any assistance. Martha and Gerald were clearly exasperated at the lack of assistance and considered this to be out of character for the region, or perhaps an indication that things were changing.

From Martha's perspective, Queensland Parks simply does not have the resources to help fight fires in other places, as they do not have the resources to control fires within the parks themselves. She reflected that each park may have an adequate number of workers during normal times, but during a crisis they lack the workforce to manage it successfully. The argument that Queensland Parks lacks the staff for the bushfire season is related to a broader sense that National Parks are inappropriate for the region, and that land was better managed when it was all cattle stations. Indeed, when destructive bushfires did descend on Rinyirru National Park in late November and early December, the park was—to use ranger in charge Ray's phrase—"skinny on staff."

In this way, wildfires function as a way for graziers, in particular, to critique land-tenure changes in the region. Wildfires reveal land-tenure boundaries as arbitrary, and left unchecked, they sweep across pastoral lease, National Park, and Aboriginal freehold blocks with impunity. To graziers, the types of land and fire management currently employed on these different forms of land tenure leave them more vulnerable to fires than they were when each block was pastoral lease and all land-holders were equally concerned with preserving pasture. Additionally, Queensland Parks and Aboriginal ranger groups' preoccupation with workplace health and safety is understood by graziers to be an impediment to successfully and efficiently fighting fires—an example of bureaucracy overruling logic.

What emerges from the way that groups of land managers frame their concerns around wildfires is that what is at stake for each group differs. For Queensland Parks and Aboriginal rangers, their primary concern with wildfires seems to be protecting infrastructure. Graziers, however, are primarily concerned with preserving pasture. If cattle stations are burned black, they lose the pasture that they rely on to feed their cattle at least until the monsoon begins. With wildfires sometimes starting months before the wet season (as was the case with the fire at Marina Plains), graziers potentially face months of attempting to support their cattle on significantly reduced pasture. From this emerges the sense among graziers that they are particularly vulnerable to fire, and this is reflected in their responses to wildfire vis-à-vis the less urgent responses from Aboriginal Land Trust and Queensland Parks rangers.

Material interests emerge as an important site of tension between different land managers' responses to wildfires. While Queensland Parks and Lama Lama rangers were primarily concerned with protecting their infrastructure in the example mentioned earlier, Alan was concerned with protecting his pasture. Similarly, where Martha and Gerald sought to control the fire on their station, this fire presented less of a threat to their neighbors, and, accordingly, they did not provide assistance. These divergent concerns reflect the different threats that fire represents across a range of land tenure types. Queensland Parks rangers may consider it to be deeply unfortunate if large swathes of land are burned in the park, but their main concern is to protect Queensland Parks infrastructure that in some cases is protected as cultural heritage and in other cases is expensive to repair or replace. For graziers, there is more that is immediately at stake when wildfire spreads through paddocks, as they rely on pasture to feed their cattle and thus rely on the existence of grasses in order to maintain their livelihoods. On Aboriginal land, Aboriginal Traditional Owners and rangers are concerned to protect infrastructure, important story places, and, in some cases, pasture. Importantly, out-of-

A ranger observes backburning for a bushfire, Rinyirru National Park, 2020.

Ash and smoke from bushfires in the storm winds, Rinyirru National Park, 2020.

control wildfires can also affect how lucrative engagement in the carbon sequestration scheme is on both pastoral lease and Aboriginal freehold. If late dry-season fires spread, uncontrolled, land managers may miss out on any carbon credits payments. Thus, economic relations to land, and more specifically the various forms of land tenure and property in Cape York, shape how different people respond to wildfires.

Wildfires draw different land managers into relationships, as people work with or across each other to protect their material interests. The wildfires described here both generate and illuminate frictions by bringing to the fore different economic relationships to land. Along with carbon credits, cool burns, and storm burning, wildfires can function for land managers as a lens through which to critique how other people care for their land—by either burning too much or too little, or by showing a lack of regard for their neighbors in how they manage wildfires. In this way, talk and practices around fire operate as a way for land managers to position their own fire knowledge and forms of land management as superior to that of their neighbors, even as burning practices are intimately linked with economic relations to land.

––––––––––

Fire in its various forms structures land management and involves different values around land and how people understand the human role in managing landscapes. Despite fire management regimes looking quite similar from land manager to land manager, these burning practices emphasize a variety of things. For Aboriginal Traditional Owners, fire management is a cultural activity, whereas for graziers burning is about preserving pasture and mimicking what they understand to be a natural or prehuman burning regime. These relationships to fire differ from those of Queensland Parks rangers, who base their fire management on Western science and hazard reduction. Aboriginal burning has been adapted and transformed through the intercultural spaces that existed on cattle stations. In Cape York, it is now difficult and, indeed, counterproductive, to demarcate Aboriginal fire management from cattle station fire management. The kind of burning regime employed today on Aboriginal land, pastoral leases, and National Parks is modeled on the burning that was carried out by Aboriginal stock workers and graziers on cattle stations. In this "zone" of interaction (Merlan 2005), such burning likely underwent adaptation and change and served multiple uses for different people simultaneously. Shared contemporary burning practices, then, are the result of the intercultural co-creation of environmental knowledge.

Fire management can be understood as a site of tension and interaction in which the interacting of different groups of people, different types of fire knowledge, and the government-initiated carbon sequestration program give rise to new types of burning and new forms of critique. Graziers critique the tendency of Queensland Parks and some Aboriginal groups in Cape York to exclusively implement cool burns, suggesting that storm burns are needed in order to emulate a "natural" (prehuman) fire regime. This is related to the idea among graziers that savanna landscapes should be open. Entwined with this notion is the reality that an open savanna landscape is more advantageous for running cattle. Grazier burning is, then, largely concerned with protecting and preserving pasture, reflecting the value graziers place in their own labor and a workable landscape.

It is through historical and contemporary forms of everyday interaction that contemporary burning regimes in Cape York emerged and persist. Such intercultural interactions allowed for the sharing and co-creation of fire-related knowledge. However, the divergent purposes that seemingly similar burning regimes are geared toward give rise to tensions between groups of land managers. The interaction between different groups of people, different fire ontologies, and a government-led response to climate change through the carbon sequestration program precipitate new types of burning and new forms of critique. Lighting and fighting fires can be understood as an act of care for Country, the land, and livelihoods, or, when done as a way that is perceived as self-serving, can be framed as exhibiting a lack of care—for the land and for one's neighbors.

Inundation at Hann Crossing, Rinyirru National Park, 2019.

SIX

Water

It was a hot and muggy day in October, and the air had that charged quality that might mean rain was coming. I was mustering with cattle grazier Alan. We sat in his Toyota, which he sometimes used to lead a herd of cattle while his sons and wife worked on quad bikes and a pack of well-trained dogs kept the sides and tail of the herd in check. There was a wide blue sky, pale and washed out, and creeping heat. The landscape looked bleached, baked colorless by the sun. By this point in Cape York's dry season, there was only minimal groundcover remaining, composed of clumps of dried-out, brittle, silver grasses. We sat, waiting for Alan's family to move the mob of cattle a little way. The sun was beating down on the car, with a breeze blowing hot from the northwest. Alan scanned the sky as we waited. "You start looking out for any sign that the wet is coming soon," he said. He told me that there are normally storms in Weipa, on the west coast, about a month before they get rain here in the east. "No storms in Weipa yet," Alan said with a wry smile.

Speculation about when the wet season will begin dominates conversations from October onward. By the end of the dry season, drought conditions can make life difficult for graziers. They must deliver nutritional supplements (lick) to their cattle, cart water, and check on their dams, keeping a close eye on the condition of their cattle. Graziers listen to the radio, waiting for reports of storms on the west coast. As well as monitoring weather forecasts, graziers follow Aboriginal Traditional Owners in observing environmental indicators. They listen for the call of the storm bird and see the presence of blue tongue lizards as an indication that the wet is coming. Operational Queensland Parks rangers also follow weather reports closely as they, along with off-park management, must decide when to close the park to visitors for the wet season.

When the rain comes, the change is rapid and transformative. The first storms and rain bring an explosion of vegetation and growth and douse the late dry season fires that Cape York inhabitants have been cautiously monitoring. In 2018, the first big rainfall came in early December. I had driven north to the community of Yintjingga through a dry and dusty landscape, but when I returned south several days later, I found everything covered in a carpet of startling bright green vegetation. By January, the entire landscape was transformed. Rivers that had been dry only a handful of weeks before were flowing and vegetation was growing rapidly. Roads became muddy and impassable, and the rhythm of life for Cape York inhabitants shifted.

The wet and dry seasons function as cyclical agents that precipitate different modes of living and working among people in Cape York. Water, in both its scarcity and in its abundance, mediates and transforms life and work. It has an immense impact on the lives of people in the region—as drought, as build up, as monsoon, as cyclone, as flood. It divides the year. Water in Australia's north is temporal; it is rhythmic. And, importantly, speculation about the onset, duration, and amount of rainfall of the wet season frequently bleeds into discussions around climate change.

Working closely with the land and relying on the wet season to ensure the well-being of waterways and pastures and extinguish wildfires means that land managers in Cape York are keenly aware of the climate variability they experience in their region. While close observation of the weather and climate is shared by graziers, Aboriginal Traditional Owners, and rangers alike, these groups understand and respond to the concept of anthropogenic climate change in diverse ways. Depending on their economic and social location, people express a diversity of views on climate change, spanning rejection, acceptance, and ambivalence. For some land managers, the explanatory model of natural cycles is more convincing, allowing them to observe climate variation without couching it in the language of "climate change." Intercultural relationships lead people to consider the notion of climate change in different ways. Importantly, the explanatory models that different people draw on to understand climate variability are entwined with their livelihoods. Taking attitudes toward the concept of climate change as my starting point, in this chapter I think through how a seemingly uncaring position is undercut by and through caring labor, exploring the gap between a stated politics and deeply held environmental values.

Changes to Life and Work

Due to its tropical climate, Cape York is characterized by an intense seasonality. I follow Krause in thinking of these seasons as rhythms, rather than "temporal blocks" ' (2013, 42). Such an approach allows an interrogation of the way in which shifts in dwelling and activities occur along a spectrum and an understanding of "a social and ecological world in motion" (Krause 2013, 42). Life and work are transformed for Cape York land managers during the wet season, although when exactly and the extent to which this happens varies significantly from year to year.

I spent much of the early wet season in 2019 stranded in the city of Cairns. With road closures and flooded bridges, I was unable to visit most of my research participants once the monsoon had well and truly set in. However, many Aboriginal Traditional Owners from the Cape relocated to Cairns for some of the wet season, and I spent time with the extended families of people I knew—catching up in the shopping center food court, ferrying children to and from water holes, and visiting some older people in their temporary accommodation adjacent to the public hospital. Having resigned myself to the fact that my mobility was going to be greatly reduced over this time, I had organized to spend several weeks at the now water-bound Rinyirru National Park. Remote national parks—including Rinyirru National Park—in the region are closed to tourists during this time, with the timing of park closures determined by local government authorities. These closures instigate a significant shift for the rangers who work there.

Eager to see what life was like for the rangers stuck in this remote park, I booked a seat on the mail plane, which makes its weekly rounds up and down the Cape. As well as me, the small ten-seater plane was occupied by the pilot, a copilot who seemed to also function as the mail worker, and a handful of Cape York locals who had traveled to Cairns to receive health care. Seats on the mail plane were surprisingly inexpensive, a fact I later learned was due to their being subsidized by north Queensland (rugby league) footballer Jonathon Thurston. As I took my seat on the plane, I was handed a small plastic bag containing some snacks and a water bottle emblazoned with Jonathon's face and text advising me to "get [my] Thurst on." As the plane flew north, I spotted particular landmarks that I recognized—towns, roadhouses—and was amazed to be able to see the quantity of water that covered much of the landscape. Depressions in the geography were now underwater, giving the impression of a vast, unending wetland dotted with little islands of trees.

Once in the park, I was greeted by the rangers and felt a festive atmosphere, a significant shift from my previous visit during the wildfire season. I was shown to a vacant donga—a small, prefabricated house on a concrete slab—and set myself up for a month in the park without mobile phone reception or internet access. The rangers who live in the park have access to connectivity, but it is something they must arrange and pay for themselves. Queensland Parks rangers remain working throughout the wet season, yet various flooded roads make it impossible for them to leave the park for weeks or even months at a time. The wet is considered to be a time to complete as many tasks as possible, without the distraction of the public to grapple with. During the wet season, the normal nine days on, five days off roster of the dry season is replaced by a regular Monday to Friday working week in which the rangers attend to tasks like spraying weeds, maintaining or repairing infrastructure, mowing around the ranger base (this in itself is almost a full-time job), water monitoring, and tree planting. It was during this period that I sprayed the seemingly endless patches of lion's tail that I discussed in chapter four. The wet is also a time in which the rangers are able to engage in enjoyable activities on the park, visiting different waterholes and rivers and fishing recreationally. We spent weekends piling into the amphibious Argos, an uncomfortable vehicle that can traverse land and water (though at some cost, I might add), in order to access different parts of the park.

At this time of year, only Queensland Parks rangers were present in the park. The Aboriginal Land Trust rangers who are employed by Rinyirru Aboriginal Corporation work only until the end of November and return home for the wet season, to Hopevale, Cairns, Kuranda, or elsewhere. Rinyirru Aboriginal Corporation does not have the funding to employ their rangers over the wet season, and during this period there is also less work to be done in the park. Land Trust rangers either find short-term employment to cover these months or live without an income for four to five months. Some decide not to return to Land Trust work in the new year. For instance, Eddie is a Land Trust ranger who often works picking bananas during the wet season. He had done this kind of seasonal work for the same company for many years and told me he was confident that he would be able to secure work during the park closure. Another ranger, Donna, returned to Hopevale during this time and engaged in short stints of other work, like making naturally dyed silk scarves with the Art Centre, or harvesting seeds from native plants to sell. When I visited her toward the end of the wet season, Donna and her partner were camping out in an old house belonging to his family, not too far from the beach near Hopevale. The house, having been empty for some years, had no electricity or running water, though there was a small creek at the bottom of the yard that was possible to bathe in. Donna was spending her time driving in and out of town, caring for various family members, and helping her partner settle back into the community after a period of incarceration. Together we drove down the beach and walked among washed-up rubbish on the sand as Donna explained that she regularly came here in search of items of value that she could use or sell on.

In Lama Lama National Park to the north, the Aboriginal Lama Lama rangers also go on a break from work during the wet season. They finish their year in early December, with full-time rangers usually returning to work in February and casual rangers returning in May or June. In early December, I visited the Lama Lama ranger base in Port Stewart/Yintjingga in the hope of spending time with some of the Lama Lama rangers but found that most had already shifted out for the wet season. Some older people had told me that they remained at Port Stewart/Yintjingga through the wet season, leaving only if they had to evacuate for a cyclone. However, the reality was that younger relatives would take older people with them into town—either Coen or Cairns—

because the risk of their being cut off by rain was too great. There is only one road between Port Stewart/Yintjingga and Coen, and it requires crossing several rivers, many of which become impassable when it rains. While several young people in their late teens and twenties live more or less full-time at Port Stewart/Yintjingga, during the dry season they may make multiple trips each week to the nearby town of Coen to socialize and purchase groceries, cigarettes, and alcohol. Additionally, the low-lying dwellings at Port Stewart/Yintjingga are at risk of flooding from heavy rainfall events and cyclones.

For cattle graziers, work remains constant but shifts into a different register. Graziers use the wet season to complete administrative tasks and fix things around the station and their houses, although the wet season also provides the opportunity to socialize with other graziers. One of the main work activities that graziers take part in during the wet is maintaining and repairing fence lines. In the wake of any substantial storm that may have caused branches to drop or trees to fall, or any flooding, fences must be checked. Maintaining boundary fences between pastoral leases and other types of land tenure takes primacy. During the wet season, graziers' cattle frequently wander into neighboring national parks as fences come down and floodgates are washed away. Floodgates are a difficult type of fence to maintain. Many riverbeds are twenty or more meters wide, and thus wire must stretch from bank to bank—often with pieces of corrugated iron dangling down to try and deter cattle from ducking under the wire in deeper sections. Graziers accept that their floodgates will inevitably wash away with each flood, and while repairs are considered pressing, they are sometimes delayed by floodwaters, flowing creeks, and boggy sections of soil. Water in this instance "does something" (Hastrup 2014, 23) in the social realm as it sweeps away fences and floodgates, allowing cattle to wander into national parks. The flow of water, here, contributes to ongoing tensions between graziers and Queensland Parks. Cattle move through once-standing fences to better pastures, pulling graziers into relationships with the complex bureaucracy of Queensland Parks as graziers must then apply for a permit to muster their cattle back.

In Cape York, as in other places, water represents both threat and survival. Water can symbolize a threat when it is at either end of the spectrum, scarcity in drought or destructive oversupply in flooding.

Water also represents survival, and the monsoon is necessary to replenish the pastures, dams, and river systems that land managers in Cape York rely on and care for in different ways. The terminology around different seasons in northern Australia—the "wet" and the "dry"—refers to the behavior of water, dividing the year into periods of less and more water. As Veronica Strang notes, water is a constantly transforming element, shifting between "oppositional extremes" (2004b, 49). Strang writes that water "may be a roaring flood, or a still pool, invisible and transparent, or reflective and impenetrable" (2004b, 49). Because of this, Strang argues, water lends itself to multiple meanings and metaphors. Strang suggests that "thinking with water" can result in analyses that attend to "shifting and mutually constitutive processes" that encompass relations between humans and nonhuman forces (2013, 186).

In Cape York, by "thinking with water" it is possible to see how water directly shapes people's livelihoods and affects where people dwell throughout the year. In the case of the community at Port Stewart/Yintjingga, water has ramifications for social relations as it brings generational differences into tension, revealing divergent aspirations and forms of belonging between older and younger Lama Lama people, as younger people take older relatives into town with them for the duration of the monsoon. For graziers, water enables their stock to wander and pulls them into (frequently tense) relationships with neighboring national parks that seek to exclude cattle. Hastrup (2014) also advocates for considering water's agentive powers. As she notes, "water *does* something in society. Water irrigates, inundates, floods, dries up, and creates social tensions as well as transport systems" (Hastrup 2014, 23, emphasis in original). Of course, both the presence and absence of water affects people in diverse ways owing to complex intersections of inequality, whether these impacts are related to a lack of water in drought (Jorgensen 2016; Roberts 2019), an oversupply of water in flooding (Casagrande, McIlvaine-Newsad, and Jones 2015; van Voorst 2014; Bankoff 2003), or access to subterranean water resources (Bessire 2021; Babidge 2019).

In Cape York, for instance, cattle graziers experience drought as a stressor because of its impact on both pastures and water sources; Aboriginal Traditional Owners living on remote outstations experience flooding as a stressor, as it necessitates evacuation from their homes for the duration of the wet season. Water can operate as a "theory machine"

A flooded forest, Rinyirru National Park, 2018.

Travelling on the Argo to go fishing, Rinyirru National Park, 2019.

(Hastrup 2014, 27), through which we can find new ways to understand social relationships around water use, control, and access. The monsoon "does something" in Cape York—with impacts ranging from how and where people are able to work to where people dwell during the wet season. At times, the impacts of the monsoon are more rapid, intense, and dangerous, such as when tropical cyclones hit the region.

Enduring Extreme Weather

Far North Queensland is subjected to tropical cyclones relatively frequently. In late March 2019, Cyclone Trevor hit Cape York and created a large rain event causing significant damage across the peninsula. Up to this point, the wet season had brought reasonably good rainfall, although everyone had been holding out for a little more to see them through the dry season. Cyclones had been a vague threat, coming and going for months but losing momentum by the time they made landfall. Over lunch in a Cairns shopping center food court in early March some Lama Lama women had joked to me about a "zombie" cyclone that kept threatening to return. Cyclone Trevor came across the town of Lockhart River as a category three system but was later downgraded to a category one storm. Lockhart River was hit by wind gusts of 133 kilometers (82 miles) per hour and received 300 millimeters (11.8 inches) of rain in just twenty-four hours. The towns Lockhart River and Coen were left without power for days (ABC News 2019). Coen was cut off from food deliveries.

After the waters receded and the roads reopened, I ran into Lama Lama ranger Mabel at a motel in Cooktown in April. She told me that Coen had been without food deliveries since the cyclone had hit almost a month earlier. She said that there was no food in the store, no cash in the ATM or post office, and no cigarettes. I asked her what people did in this situation and she replied with a shrug that they just "stress it out." A couple of weeks later I saw another Lama Lama woman, Leena. She told me that her extended family had fared well during this period as her uncle was able to get fresh "killers" (cattle bred for consumption) from the bottom of the range near Coen. Families with access to four-wheel-drive vehicles were able to travel south to larger towns when the roads were open and stock up on supplies, but this is an expensive and time-consuming option, and at times—if the rivers have risen—is not possible even with appropriate vehicles. In such instances, local people must rely on their kin networks rather than the government to provide disaster relief and ensure they have enough food to eat. It is through these relationships, and their own ability to access resources from the land, that people are able to make do during long periods of flooded roads and little (if any) government support in the wake of increasingly frequent cyclone events.

The rain from Cyclone Trevor caused substantial flooding at the outstation community of Port Stewart/Yintjingga where many Lama Lama families spend the dry season, damaging the road between Coen and Port Stewart so that it was impassable until repairs could be conducted. The week after the road reopened, I visited Port Stewart with Leena and another Lama Lama woman. Leena warned me that there had been significant damage to the houses there, particularly in the area called Bottom Camp where she normally lived with her grandparents. The road—while now passable—still has big washouts and was a little treacherous. As we drove to Port Stewart, Leena recalled her first visit since the road reopened a week earlier. She had taken a carload of older relatives and children with her. She says that as soon as their car had approached the mountain range outside of Coen, everyone's spirits were uplifted, and the children were excited. She told me that the children "hardly left the saltwater all day."

We drove to Bottom Camp to assess the damage around Leena's grandparents' house. Sand and debris were spread across the camp and cars were pushed into holes, half-submerged in the sand. The floor of Leena's house was coated in several centimeters of mud, and the outdoor kitchen where her grandfather normally cooked had been washed away. Despite the damage to Bottom Camp, Leena reflected that the rain had cleaned the river, clearing it of debris. She also said that the shape of the river mouth at the beach had been transformed due to the large volume of water. To Leena, the river mouth was better this way and resembled how it used to look in the past, when she was a child. As well as representing threat, to Leena the flooding had also resulted in a kind of renewal. She described the floodwaters as "cleaning up" the river, even as they deposited sediment and debris throughout her house.

To Leena, Mabel, and the other Lama Lama women, cyclone events such as this are considered a normal part of living in Cape York. While Cape York Natural Resource Management (CYNRM) employees and some senior Queensland Parks management staff discussed the increased frequency and intensity of cyclone events as indicative of a changing climate, such a perspective was not articulated by these Lama Lama women. Cyclones and flooding are things to be expected and endured, and, as Leena suggests, can even be understood as restorative for the landscape. Perceptions that flooding can be positive because of its ability to "clean" waterways have also been recorded among Aboriginal people in the Kimberley region of Western Australia (Toussaint et al. 2001, 55; Toussaint 2008).

Nearby, Tidewater Station also experienced some flood damage from the cyclone. Alan and Bev, the graziers who live there, were elsewhere at the time but had a caretaker looking after the property in their absence. When the cyclone passed over, a few inches of water came through the house. This had happened before, and the caretaker was able to avoid too much damage inside the house. The shed near the house had significant amounts of water come through—a bike floated away, as did some pallets and empty woven plastic bags which caught on fences. Once the waters receded, the shed was filled with leaves, sticks, and mud, but the damage was minimal. After the flood it was some time before Alan and Bev could return home. The cyclone had damaged the Peninsula Development Road, and the government department responsible for the road had opted to close the road until the damage could be assessed. I spoke with Bev on the phone shortly after the flood and she told me, laughing, that everyone had been hoping for a little more rain before the wet season finished but this was more than they bargained for. "Oh well," she reflected, "you can't pick and choose."

Later, when I visited Tidewater Station, Alan pointed out that the marks on the termite mounds showed how high the water level had been during the flood. Tidewater is low-lying, and when floodwaters are high enough to reach the house there is little high ground left for the cattle to shelter. Alan was concerned that they had lost some of their younger calves in the flood. He said that they usually did. While I was there, he received a phone call from CYNRM, who were trying to determine whether Alan and Bev needed disaster relief funding. Once his phone call ended, Alan told me that it was too difficult to quantify their losses.

They had no real way of telling how many calves they had lost, and no precise sense of how many kilometers of fence line required repair. "I wish I could lie, but I just can't," he said.

Alan and Bev's decision to refuse disaster relief funding is tied into a kind of ambiguity that Bev highlighted. She pointed out that it is difficult to draw a line between what is normal wet season flooding and what is extreme weather flooding. They flood every year, but some years are worse than others. Bev said that it is difficult to know how high the water ought to be before they are eligible for assistance. Putting aside Alan's claim that he does not believe in "handouts" or government assistance anyway, this ambiguity around what is normal weather and what is extreme weather is significant for how some Cape York residents frame the wet season, cyclones, and experiences of climate change.

Since the 1980s, disasters have come to be understood in the social sciences as both a fundamental element of environments and, importantly, as exacerbated or even created by human action (Hoffman and Oliver-Smith 1999, 2). Around the disaster itself is a nexus of social constructs that shape and frame the disaster in certain ways. What is considered to be a "disaster" rather than a mundane event is culturally contingent, but often within societies is defined by a particular powerful authority. Indeed, a common narrative around disasters is that they are "non-routine," disruptive, and destabilizing. Thus "disaster" is held up as the polar opposite of ordinary life (Oliver-Smith 1999, 23). Such framings of disaster are problematic because they rest on an assumption of societal order before the disaster and assume disasters to be the result of natural hazards—when in reality many "natural" hazards would not become disasters with the right kind of societal management (Oliver-Smith 1999, 23). The existence of disasters, then, is the result of what Oliver-Smith calls "adaptive-failure" (1999, 29).

Cyclones and flooding are particularly ambiguous in Cape York, because they are a routinized part of life in the wet season. As Bev pointed out though, some years are worse than others, and it is difficult for land managers to really determine the line between routine and extreme weather. In the natural sciences these weather events are defined as "natural disturbances" rather than "natural disasters" as they exist well within the parameters of what counts as "normal" weather for Cape York (Turton 2008; Lin, Hogan, and Chang 2020). Further, because of the relative self-sufficiency that remote living already requires—such that, even when residents must simply "stress it out" to endure periods of isolation—people are able to access food through networks of kin or their own hunting activities. As in other places that experience extreme weather and flooding events (Casagrande, McIlvaine-Newsad, and Jones 2015), kin and peer networks, rather than official emergency services, are vital for both the preparation stages and in the immediate aftermath of an event. It is, perhaps, the resilience of family and kin networks such as these that allows people in Cape York to consider cyclone events as expected and manageable, even when government support is lacking or absent. Yet, these "normal" weather events are frequently destructive, and families and communities often require funding assistance for extensive repairs in the wake of such weather events, and the role of government assistance tends to become more important for longer-term recovery (Casagrande, McIlvaine-Newsad, and Jones 2015, 359). The framing of these kinds of weather events as "normal" is related to the kinds of environmental narratives (Olwig and Rasmussen 2015) that prevail among locals in the region. Dominant environmental narratives can shape how people come to understand the norms for weather in a particular region. In Cape York, the notion that cyclones and flooding events are within the "normal" range of weather events masks the changes that are occurring in terms of the frequency and severity of these disturbances.

This ambiguity is reinscribed in the paradox of a lack of government sanctioned disaster declaration over the entire region, despite government funding dedicated to disaster relief being available, administered through CYNRM on a needs basis. This leaves the CYNRM practitioner along with landholders to decide whether damage has been substantial enough to warrant a consensus that a disaster has occurred. Cyclones and flooding are an expected and accepted part of the wet season for graziers, Aboriginal Traditional Owners, and Queensland Parks rangers alike, even if their impacts vary from mildly inconvenient to dangerous and destructive. For land managers in Cape York, understandings of climate variation are shaped by peoples' economic relationships to land and views on what constitutes a "green" political sensibility. Where organizations like CYNRM and Queensland Parks interpret increasingly severe cyclones and flooding events as evidence of a globally changing climate, other land managers, like graziers, understand variations as linked to long-term natural cycles.

Framings of Climate Change

In early June 2018 I visited some large salt pans in the southern section of Lama Lama country with a group of Lama Lama rangers. We walked along a crevice where mangroves grew and stopped at various metal posts indicating climate change monitoring points to take photographs of the vegetation. These photographs were mostly being taken by a non-Aboriginal scientist, Helen, who had worked with the Lama Lama rangers for several years. As Helen took the photographs, senior Lama Lama ranger Patricia explained to me that the tides come as far inland as these mangrove trees. Patricia said that when she was a child in the 1980s, she used to camp at the creek near these salt pans with her old people, spending time fishing, walking the Country and learning stories and language. She said that the trees back then used to be greener. Now, she reflected, it's dry and many trees are dead or dying. She attributed this to climate change. She listed a range of other impacts of a changing climate to me, mentioning coral bleaching on the reef in Princess Charlotte Bay that saddened the older community members, and a particular swamp that used to be freshwater turning brackish.

Months later, on a visit to Silver Plains Station on another part of Lama Lama Country, Patricia and I sat at the kitchen table in the old house there, drinking tea with Lama Lama elder Charlie. Charlie and Patricia were discussing issues with the groundwater supply at Silver Plains. The water level in the spring that supplied the houses at Silver Plains with drinking water had become dangerously low. Charlie considered this to be something that happens from time to time and referenced incidents in the past when the groundwater supply had been more and less plentiful. He expressed a faith that the spring would not dry up completely, based on anecdotal evidence from his ancestors about the spring's permanence. Patricia disagreed with Charlie, contending that the groundwater supply was being affected by climate change. She began to describe other climate change impacts that the Lama Lama rangers must grapple with. The threat of coastal erosion from rising tides is particularly acute. Patricia spoke about the risk rising tides present to rock paintings on Marrpa Island and Cliff Island—part of Lama Lama Country. She told me that the paintings were dated at 30,000 years old, but that the tides were already reaching the cliff faces where they are located and were starting to erode and damage the paintings. She, along with other Lama Lama rangers, visited these islands relatively regularly for different environmental surveys and had noticed how over only a handful of years erosion had affected the paintings.

Lama Lama people are experiencing firsthand the impacts of a changing climate. They are reading changes in the landscape occurring within their own lifetimes. The most obvious and noted of these is rising tides and saltwater incursion. However, the tendency to link these changes to anthropogenic climate change is a result of the interpenetration of a Western scientific land management model into an Aboriginal land management model. Through their observations and monitoring of changes in their homelands, Lama Lama people are engaging in a kind of "ethnoclimatology" similar to that described by Crate (2008).

For many of the Indigenous and local peoples around the world whom anthropologists work with, climate change represents a threat to subsistence, access to resources, livelihoods, and the physical world more broadly, as well as potentially destabilizing Indigenous cosmologies (Crate 2008; Connor 2016). Yet, as Potawatomi scholar Kyle Whyte (2017; 2018) points out, the discourse of climate change is not experienced homogenously by people in climate-affected places. For Whyte, climate change is not framed as heralding of unprecedented times; instead, it is understood as a continuation of the destruction wrought by the ongoing unfolding of colonialism. Climate change is understood by many Indigenous peoples around the world as directly traceable to colonization itself, and the exploitative relationship to natural resources that colonialism fosters and supports (Callison 2014). Having survived and endured multiple iterations of colonization (including invasion of territories, frontier warfare, missionization and forced removal, and loss of autonomy over labor power as people were pulled into cattle industry work), Aboriginal people in Cape York are now observing further fractures and changes to their lifeworlds and homelands wrought by climate change.

While Lama Lama people may have already been observing changes in the landscape of their homelands, it is through interactions with visiting scientists like Helen and their ongoing engagement with Queensland

Parks that they have come to frame these changes as linked to anthropogenic climate change. Across the world, in regions where the impacts of climate change are observable on a local scale, local people are engaging with scientific discourses of climate change. As Cruikshank notes, "if climate change is indeed global, its consequences are profoundly local" (2005, 25). The global and local focus on climate change has precipitated a space of encounters between Indigenous Peoples and Western scientists. Where Cruikshank (2005, 255–56) points to these kinds of relationships as sometimes leading to the exclusion of indigenous peoples from managing their lands, in the case of Lama Lama rangers the opposite is true as a result of the land tenure arrangements in Cape York. Lama Lama rangers are supported by state and federal government funding and the management frameworks co-created with Queensland Parks to continue to monitor the impacts of climate change in their homelands. The focus on climate change monitoring has, then, helped to bolster the funding that the Lama Lama rangers receive, allowing them to maintain and even expand their workforce. To Lama Lama people, ideas around climate change are relevant to their day-to-day lives and concerns for their livelihoods, land, and material culture. Further, their embrace of climate change discourse demonstrates an ability to take advantage of government priorities in order to maintain a degree of control about the knowledge and research that is produced on and about their land. However, as Callison observed among her Inuit interlocutors in the Arctic, Indigenous people are often "much more able and eager to integrate scientific findings into their own systematic observations than scientists would be in their encounters with [Traditional Knowledge]" (2014, 61).

Couching environmental changes in the language of climate change was prevalent among the rangers in Rinyirru National Park—both Rinyirru Aboriginal Corporation rangers and Queensland Parks rangers. Queensland Parks rangers tended to associate the extreme rainfall events and flooding of early 2019 with climate change, particularly in relation to a significant flood event in northwest Queensland that occurred in February 2019 (Smee 2019). Some Queensland Parks rangers spoke about the size of wet seasons in relation to the El Niño–Southern Oscillation (ENSO), displaying a basic knowledge of the patterns inherent in ENSO and pointing toward the tendency of Rinyirru National Park to experience substantial flooding every six or seven years.

Comfort with the concept of anthropogenic climate change led rangers to interpret large flood events as linked to human actions. One afternoon during the wet season, the rangers were discussing the recent flooding in northwest Queensland. Ray, the ranger in charge, reflected that "Mother Nature seems pretty angry," to which another ranger, Ivan, replied, "well, she's probably got a few things to be angry about." While such a personification of Mother Nature was deployed in jest, and greeted with laughter from the rangers present, it reveals how Queensland Parks rangers are comfortable with the Western scientific framework of attributing climate change to anthropogenic causes. The kind of land management carried out by these rangers is largely based on a Western scientific model of environmental management, and thus, the concept of climate change fits into this system for seeing, interpreting, and managing landscapes. While these land managers have varying levels of knowledge about the science of anthropogenic climate change, it is a concept they take for granted as important for their work on some level.

In a slightly different vein, several Aboriginal Traditional Owners interpreted the significant rain events and longer wet season as a sign that ancestral spirits were displeased with the actions of some Aboriginal people in affected areas. Peter, the senior ranger for Rinyirru Aboriginal Corporation, told me that he and his uncle, a respected elder, believed that ancestral spirits had created the rain because they were unhappy with how some families were working in the national park. In particular, Peter felt that some people who had gained employment through Rinyirru Aboriginal Corporation were improperly claiming their ancestral connections to the park in order to secure employment or hold more sway in terms of management decisions. To Peter, then, extreme weather events could be directly attributed to the actions of humans, but in a much more specific way than the concept of anthropogenic climate change embraced by the Queensland Parks rangers. Yet Peter, along with other Rinyirru Aboriginal Corporation rangers, also discussed the impacts of anthropogenic climate change on Rinyirru National Park.

Interpreting the actions of individual Traditional Owners as affecting the weather is, to Peter, not in conflict with the concept of climate change. Instead, Peter's perspective indicates a kind of syncretism of a belief in the power of ancestral spirits and acceptance of the Western scientific model of climate change. Peter is able to interpret extreme weather events as partially caused by the transgressions of Traditional Owners angering ancestral spirits, and partially as the result of anthropogenic climate change. He, along with other Traditional Owners, mobilizes the explanatory framework of climate change where appropriate, and at other times draws on alternative models. Our conversation about flooding in the park came at the tail end of a discussion about cultural protocols not being correctly observed by some Aboriginal Traditional Owners working there. Whereas in other contexts, like a discussion about coastline erosion with an employee of Carpentaria Land Council, Peter instead drew on the discourse of climate change. This kind of interpenetration of different knowledge systems has been referred to by Stensrud as "producing the entanglement of worlds"(2016, 86). In the "entanglement of worlds" that the joint management context gives rise to, Peter mobilizes different knowledge systems and explanatory models depending on the context and does not appear to experience these knowledge systems as being in conflict.

Climate knowledge is always historically and socially contextual and garnered from an aggregate of multiple sources (Hastrup 2016; Crate and Nuttall 2016). Kirsten Hastrup (2016, 39–40) advocates for tracing the genealogy of the concept of climate change, from what she calls "a scientific puzzle" to a "more ominous and heavily politicized arena of discussion" and becoming a "fact" with the establishment of the Intergovernmental Panel on Climate Change (IPCC). In Hastrup's (2016) research among northwest Greenland communities, the language of climate change has become prevalent. As well as observing and experiencing firsthand changes to their lived environment and the impact this has on lifestyles and subsistence, Hastrup (2016) found that the communities she works with are "well-versed" in the IPCC reports. As she notes, "even in this remote corner of the world—which is of course at its own center—there is no such thing as local knowledge as opposed to scientific knowledge. All of it enters into one located knowledge space, producing its own *spatial vernacular*" (Hastrup 2016, 44–45, emphasis in original).

Caring for land through the context of joint management has meant that specifically Aboriginal explanatory models for weather events and changes—like Peter's attribution of a large rainfall event to the culturally improper actions of specific Traditional Owners—sit alongside the Western scientific framework of climate change. Aboriginal rangers engage in climate change monitoring, receive funding for specific climate-change-related projects and take part in research projects spearheaded by Western-educated scientists. Thus, the language of climate change seeps into other ways of perceiving environmental shifts and in joint management, too, a "spatial vernacular" (Hastrup 2016) of interculturally produced climate-related knowledge emerges. But while many Aboriginal people in Cape York, alongside Queensland Parks rangers, are comfortable with the language and concept of climate change and are able to frame some local environmental changes as climate change impacts, the same is not true of graziers.

Natural Cycles

Toward the end of the dry season, I visited graziers Martha and Gerald at their station. One morning, as Gerald prepared to take a truckload of water to a dam that was drying out, Martha discussed rainfall with me. Like other graziers I worked with, Martha and Gerald, from time to time, had research scientists visit their lease to conduct surveys of species like tiger crabs, possums, and rock wallabies. Around fourteen years prior, Martha and Gerald established a number of monitoring points around their lease, from which they take photographs at specific times of the year to observe changes in the vegetation as well as recording rainfall. As Martha described this to me, she shuffled around some books on the kitchen table, eventually extracting two large folders where she keeps her weather observations. "I'm not sure why we do it when we're not being paid for it," she told me, laughing that she supposed she had just grown used to doing it. As well as keeping meticulous paper records, each day that it rains on her lease, Martha uploads their rainfall amounts to the Bureau of Meteorology. Many landholders in regional and remote Australia do this, although, as Martha noted to me, not all her neighbors were as diligent as she was about this. She told me that she has grown accustomed to monitoring things, and so she finds uploading the data to be straightforward and not onerous.

Reflecting on rainfall, Martha told me that a shorter, drier-than-usual wet season the previous summer meant that 2018 was a particularly dry year for them. In fact, it seemed as though their rainfall for the year would be quite similar to 2009—the driest year they had ever experienced. She reflected that, since 2009, drier years had been increasing in intensity and frequency; however, she dismissed any indication of climate change implied in this by claiming that a relative had spoken about similarly dry years in the 1940s and 1960s.

This was one occasion among many in which Martha framed environmental shifts as part of a "natural cycle" rather than an indication of climate change. Her husband, Gerald, too, reflected that the last fifteen years had been drier than the years preceding but exhibited a kind of persistent optimism and faith that things would eventually go back to how they used to be. Martha frequently pointed out that records of rainfall and temperatures "only go back so far." She believed that any shifts in climate she has experienced were part of a broader pattern that could not be properly seen because of the shallow temporal depth of recordkeeping. She told me that her grandmother had recounted stories of it being so hot that birds were falling out of the sky. Martha had never seen that happen and so comfortably believed that there were hotter temperatures in the past, before records were kept.

At another cattle station, I encountered further skepticism. A visitor made a joke about climate change, leading to a conversation between grazier Pam and myself as we washed up the dishes from morning tea. Pam told me flatly that she did not really believe in climate change. She then qualified this by saying that she believed that the climate is changing but she neither thought that it was a problem or that humans were responsible. She told me that she was aware the climate has been changing "since time began," pointing to ice ages in the distant past. Her position became more ambiguous, as she said that she thinks "overpopulation is not real good and neither is a lot of other stuff that humans do." Pam then ended the conversation abruptly by saying that she thought that climate change, along with the hole in the ozone layer, was invented by scientists to keep them in a job. Pam, along with many graziers in Cape York, harbored a deep suspicion of scientists, and thus her position on climate change was caught between a rejection of Western scientific explanations and a lifetime of environmental observation indicating that the climate is changing.

Pam here exhibits some of the same multiplicity as Peter, an "entan-

Drying out, Tidewater Station, 2018.

Saltpans, near Yintjingga, 2020.

glement of worlds" (Stensrud 2016) both similar and vastly different. She is caught between her observations of a changing climate—a result of her work on the land and close monitoring of weather conditions over many years—and a cultural value system that is, in part, predicated on a rejection of "green" politics and the expertise of urban, university-educated scientists.

While none of the graziers I worked with strongly expressed a denial of climate change, most seemed to hold mixed views. This stance, of attributing the causes of climate change to both human agency and natural cycles, is not unique to Cape York graziers and has been documented in other studies of climate change thinking among diverse social groups (Connor and Higginbotham 2013; Higginbotham, Connor, and Baker 2014). Graziers frequently reject the notion of climate change in favor of natural cycles as they discuss its effects, while at the same time acknowledging that industrial society, urbanization, and consumerism are contributing to environmental degradation. They are aware that the cattle industry, too, is implicated in emissions, though this was rarely something that they discussed.

In part, graziers' ambivalence reveals the ambiguity around what counts as markedly different climatic patterns in a place like Cape York, which has always experienced extreme weather to some extent. However, perhaps more importantly, this also suggests something of what is at stake when Cape York locals—particularly graziers—either reject, accept, or are ambivalent toward idioms of climate change. This stance provides a way for graziers to frame and understand the climatic changes they are observing, while avoiding destabilizing their own value system, within which cattle are central.

Ambivalence or skepticism toward anthropogenic climate change explanations is wide-ranging, and something environmental and social scientists have been seeking to understand. Many explanations for this rest on something called an "information deficit model," which assumes that people simply do not have access to enough clear information about climate change (Norgaard 2011, 1). In her work on perceptions around climate change in Bygdaby, Norway, Norgaard found that something else was at play. Norgaard found that her research participants were highly educated and were both informed and educated about climate change. However, there was a sense among her research participants that climate change is too big and global an issue to be dealt with at the local level, eliciting a sense of powerlessness that led to avoidance or ambivalence. Norgaard speaks about this as a kind of denial, a "condition of 'knowing and not knowing'" (2011, 60). These findings point to the limitations of an "information deficit model" as an explanatory framework for why people do not act on environmental issues (Strang 2004c; Callison 2014; Ottosson 2019). This is because an exclusive focus education and awareness, or "just the facts," negates the cultural, economic, and moral meanings that people have already imbued environmental issues with (Callison 2014, 28–29; Ottosson 2019).

Climate change discourse is an entwinement of evidence, beliefs, ideologies, and ideas. As a concept, it has been politicized and continues to be hijacked by various political and corporate agendas. It has become a "Big Story" (Daniels and Enfield cited in Hastrup 2012, 8–9). This "Big Story" of climate change leads people to read daily weather forecasts and more significant weather events as indicators of a changing climate (Hastrup 2012, 22). The opposite is true for the graziers of Cape York, who discuss these things with reference to natural cycles, couching these events within the logic that such events have happened before and will happen again, as they are governed by broader cycles that we cannot understand with the limited temporal depth of records in Cape York.

The tendency to draw on the concept of natural cycles to dismiss or divert attention away from climate change is present in other contexts among nonurban people in Australia. For instance, writing of farmers in New South Wales, Connor explains the concept of natural cycles among the people she works with:

> The core of the "natural cycles" view is the idea that weather, especially extreme weather events, is a cyclical process over shorter or longer spans of time. Humans are not responsible for the weather nor can they control it. Natural cycles proponents read long-term climate patterns through their embodied experience of local weather. Nature will "take its course." In this view, long-term cycles of climate change are affirmed, but not the human cause of current global warming. (Connor 2016, 85)

Natural cycles function to create a sense of stability in human-environment relationships through time, and through environmental change (Connor and Higginbotham 2013). Environmental changes are positioned as both historical and predictable; the environment has changed before, it will continue to change again, what we humans do has no impact on this. Connor and Higginbotham (2013) suggest that rather than fear or avoidance motivating a rejection of climate change, the belief in natural cycles emerges instead from a strongly held belief about how nature operates, based on years of firsthand environmental observations. Further, in Cape York and among the rural farmers with whom Connor and Higginbotham worked, a belief in "natural cycles" is frequently linked to proenvironmental values and a rejection of anthropocentrism.

Graziers in Cape York, like many people in rural Australia, tend to frame climate change as a belief system. Using the terms "believe" or "don't believe" to describe thoughts around climate change allows people to consider climate change to be a "subjective certainty rather than a knowledge-based observation about reality" (Connor 2016, 85). While people may observe and discuss local weather events, changes in seasonal patterns are harder to detect and are framed as "natural cycles." Seasonal changes, then, do not necessarily indicate to laypeople the existence of climate change because, as Connor points out, people frequently "cannot detect climate change signals amid the background noise of weather fluctuations" (2016, 101; Marin 2022). However, Connor suggests that laypeople do observe these kinds of changes and synthesize them as part of broader "natural cycles" (2016, 101). The natural cycles narrative can provide reassurance, allowing people to feel that there is "nothing to fear from future changes in the weather because it is all part of a pattern with good years and bad years that have to be endured" (Connor and Higginbotham 2013, 1860).

However, graziers are not blind to the effects and impacts of climate change—or to the fact that the same environmental changes they see as indicative of "natural cycles" are read by others as evidence of a changing climate. CYNRM employee Michael explained to me that he tends to speak to land managers about the impacts of climate change and possible mitigation strategies by carefully avoiding particular terminology. As he said during our interview,

I don't, sort of, mention the words "climate change" so much, probably more just how the seasons are getting . . . the seasons are changing. The wet seasons are starting later, finishing later, everyone agrees with that sort of stuff . . . we are talking about a changing climate, but as soon as you mention the words "climate change," people have different views and they just shut up about it or think you're over here somewhere [gesturing to the left with his hand] . . . But it's going to affect our fire management, if we're going to have longer, hotter dry seasons. We're going to get more chance of bigger wildfires later in the year. And if we're going to have more intense, wetter wet seasons, that can lead to more runoff and maybe more cyclone events. Yeah, so I do talk to people about that. I suppose everyone's aware of it, they've seen it happening themselves.

Graziers simultaneously reject or are ambivalent toward the concept of anthropogenic climate change while acknowledging and planning around the associated impacts and effects. What, then, is gained from such a rejection if, as Michael says, everyone tends to agree that they are experiencing environmental change on some level? For the graziers of Cape York, anthropogenic climate change exists as part of a nexus of what could loosely be termed a "green ideology." Despite holding detailed environmental knowledge and engaging in considerable conservation work, graziers shy away from identifying as "conservationists" because of the ideological connotations implied in such a word. This rejection of the "conservationist" label is related to a generalized disavowal of urban-based Western science and a prioritizing of local, experiential knowledge, garnered over lifetimes and even generations (Reardon-Smith 2021). Similarly, for the graziers, the term "climate change" brings to mind urban-based "greenies." Further, graziers are aware that the grazing industry is understood by the broader Australian public to be contributing significantly to carbon emissions. If they were to engage seriously with the concept of climate change, graziers may be forced to reckon with their own complicity in the worsening climate crisis. By being ambivalent toward the concept of climate change, graziers are able to see their role in methane emissions as a nonissue, something fabricated by urban greenies who they frame as involved in an agenda to undermine the cattle industry. A key question, then, is whether this disavowal of their complicity in a worsening climate crisis is also a refusal to care?

To say that graziers reject the explanatory model of climate change because these models implicate them and their industry in environmental harms is to simplify and flatten their relationships to land, place, and ecosystems. The refusal to be made complicit is, indeed, part of the story, but to reduce their position to this is to ignore the complex ways in which their ideas about climate change emerge out of a system of values that has an internal logic. Graziers refuse to consider the grazing industry as simply and exclusively "bad" for the environment; they point to the ways that grazing allows for a landscape that is peopled and cared for, for the way that grazing supports environmental management that doesn't rely on insecure and short-term government funding. By subscribing, instead, to the explanatory model of natural cycles, graziers are able to hold up their own proenvironmental values alongside the value they place on cattle and cattle grazing. Candis Callison notes that "science doesn't speak with the same authority everywhere, but instead operates with a distinct vernacular that has been naturalized by some and rejected by others" (2014, 37). As Michael from CYNRM described, graziers are able to incorporate some scientific findings and synthesize these with their own observations and existing land management practices in order to engage in a kind of environmental management and stewardship that does, in fact, work toward mitigating the impacts of climate change, even if the vernacular of climate change is, by necessity, left out of these conversations. By drawing on the idea of natural cycles to make sense of the changes around them, graziers have created a framework in which these things are not in tension in the ways that they might be imagined to be.

In this way, it does not follow that graziers' discomfort with the language of climate change necessarily translates into their being uncaring. In pushing back against the idea that they and their industry are complicit in a worsening climate, there is, perhaps, a failure to *care about*. However, in the actual practices and actions that graziers engage in— through their burning, their cattle management to avoid overstocking, their control of invasive plant and animal species, and their recording and, for some, uploading of meteorological observations and data to government-run databases—the tangible work of *caring for* is evident, even where the rhetoric of care is lacking.

For graziers, the concept of anthropogenic climate change is understood to exist in opposition to their livelihoods—livelihoods in which their senses of self and belonging are intimately entwined. They possess environmental knowledge that is valued as highly local, specific, and emerging from experience and observation. As an explanatory framework, the science of climate change is considered to be both abstract and a threat. It is part of a vaguely defined "green conspiracy" that for graziers involves an enfolding of animal rights activists, government oversight, and land tenure changes. Graziers understand the discourse of climate change to be threatening, even as they engage in the carbon sequestration scheme. Thus, graziers are caught between a rejection of abstract scientific knowledge, their own involvement in the climate-change mitigation industry, and their observations of climate variability. The explanatory model of "natural cycles" allows graziers to understand and categorize the environmental changes they are experiencing without having to accept an explanatory model that, to them, is not simply settled science but is also an ideology they see as a threat to their continued existence in Cape York.

For the graziers, there is an unacknowledged tension between ideology and practice, between a stated antigreen politics that *seems* uncaring, and the tangible enactment of caring labor that graziers *do* every day. This gap between caring about and caring for points out a contradiction, but it is not one that graziers find paralyzing, or even worthy of reflection. As will be evident by now, graziers are "not *all cramped up* with concerns about structure, boundaries, and the proper relationship among concepts" (Davé 2023, 72). Instead, they confound the boundaries between things and continue to labor on the land to bring about their own continued presence in the region, a goal that relies on good stewardship and ecological care even as it troubles commonsense ideas about what environmental management should look like and who environmental managers should be.

———

Seasonal shifts have a strong impact on the lives of Cape York residents. Even those Cape York residents who do not work on the land or rely on good rainfall for their livelihoods are affected, as large rain events can be destructive and reduce mobility around the region. Water, as drought and as monsoon, "does things" (Hastrup 2014) in the social realm, with impacts for how humans relate to land, the region, and each other (Strang 2004b; 2013). The monsoon shapes and mediates forms of belonging in Cape York as it necessitates both movement and stasis for different people. While the wet season is understood by Cape York residents to be generally stable and reliable with some amount of fluctuation, most people agree, to some degree, that wet seasons are changing. Aboriginal ranger groups and Queensland Parks rangers couch these changes within the language of "climate change," whereas graziers draw on the concept of "natural cycles" to explain variations.

Aboriginal ranger groups come to understand the changes they are observing in their homelands through the explanatory model of climate change as a result of the interpenetration of Aboriginal land management and a Western scientific framework. Through the intercultural context of joint management, in which ways of understanding and caring for land are continually coproduced, Aboriginal rangers have embraced the concept of climate change. Engaging with the science and industry around climate change allows Aboriginal ranger groups to play an active role in climate change research on their homelands, particularly in research about saltwater inundation, coastal erosion, and coral bleaching. Importantly, as the example of Aboriginal ranger Peter illustrates, embracing the science of climate change is not necessarily in conflict with cultural knowledge about the human and ancestral causes of weather events. Exhibiting pragmatism and a tireless capacity for adaptation, Peter—along with many of the other Rinyirru and Lama Lama rangers I worked with—deftly entwines multiple explanatory frameworks, syncretizing different knowledge systems to make sense of the environmental changes he is observing.

While Aboriginal ranger groups work to make the most of opportunities emerging from scientific and public concern about climate change, graziers perceive the concept of climate change as a kind of threat. To the graziers, the concept of "climate change" is interwoven with a number of issues and positions they understand as belonging to "greenies" and urban environmental politics. The science of climate change, as it is based on abstract and global-scale forms of knowledge, stands in contrast to the experiential and observational knowledge graziers hold about the weather. While graziers are unlikely to reject the notion of climate change outright, they tend to gravitate toward the model of "natural cycles" to explain the climatic shifts they are observing. Livelihoods, then, either through government funding or running cattle, emerge as important drivers for whether people accept, reject, or take a mixed approach to the concept of climate change.

However, despite the diverse ways in which land managers in Cape York approach the concept of climate change and respond to the extreme weather events they experience, shaped by a complex of cultural, economic, and moral values, each land manager is laboring to find ways to anticipate, mitigate, and cope with the impacts of climate variability. Whatever explanatory model they subscribe to, be it climate change or natural cycles, land managers do not dispute that they each face the shared challenges of saltwater inundation, rising tides, increasingly severe and frequent cyclone events, longer and hotter dry seasons, unpredictable wets, and a more intense bushfire season. As CYNRM employee Michael told me, "everyone agrees with that sort of stuff . . . everyone's aware of it, they've seen it happening themselves." By continuing to engage in caring labor, through their monitoring and recording of observations alongside a multitude of mitigation measures, Cape York land managers are working to make do amid a changing climate and an uncertain horizon.

The homestead, Hillview Station, 2018.

Parrots

The first time I saw a golden-shouldered parrot, or *alwal* in the Olkola language, was early on in my time in Cape York. I had gone to visit Pam at her station. Pam has been working with the parrots since the early 1990s, and many of the known nests of these endangered birds exist on her family's pastoral lease. Like many cattle stations in the Cape, Pam's family takes part in a kind of hybrid economy, earning money from road building and maintenance, taking part in the carbon credits program, running cattle, and hosting campers. Pam also takes people on bird-watching tours. On this occasion, early in the dry season in 2018, Pam mentioned to me that she might have some bird-watchers, or "birders," arriving that evening, and as such she needed to see if she could find some parrots. "You may as well come along," Pam told me. "And you're going to need to learn to ride a bike if you're going to be hanging around here."

She quickly taught me the basics of riding a quad bike. I fumbled with it at first, finding the steering pretty unintuitive and the throttle difficult to keep at a constant speed. We drove along the main road that cuts through the station, headed toward a dam. Pam was going slowly, clearly hanging back in order to keep an eye on me, but also so that she could watch out for birds. After only a short distance, Pam pulled up.

She got off her bike, so I did the same. I followed her on foot through the grass and termite mounds to where she was looking through a pair of binoculars at the canopy of a eucalyptus. Pam pointed to two baby parrots in the fork of the tree and handed me the binoculars, giving precise directions for how to spot them. I followed the trunk of the tree from its base to the fork and saw two tiny birds. They were bright green, barely distinguishable from the leaves surrounding them. I could see the hints of other colors but would never have been able to spot them without Pam's assistance. Pam told me that she had often seen the parrots around here, and once she had turned off her bike she could hear their call and was able to locate them from there.

The golden-shouldered parrot—so called for the spectacular plumage of the adult male—is an endangered species found in the savanna landscape of central Cape York Peninsula. The parrots nest in aged termite mounds and face threats of feral cats, goannas, birds of prey, and the impacts of cattle grazing and burning regimes. Despite initiatives within National Parks established to protect the parrot, many of the known nests are located on Pam's pastoral lease. Over close to thirty years, Pam has formed an intimate relationship with the birds by walking through the bush, watching, and listening. Through the cacophony of birdcalls, she can pick the call of the parrot. Pam has cultivated what Donna Haraway calls "response-ability" in this relationship, developing the art of paying attention. Her senses are now oriented toward the parrots. She has kept meticulous records of the birds throughout this time, as scientists and researchers have come and gone, seeking her assistance and expertise. Pam recalled to me that the first bird-watchers arrived in the late 1970s, but that she and her husband and his family had "always" been aware of the parrot.

In the early 1990s, two parrot scientists spent three years living at Pam's station engaging in research on the parrot. Pam assisted them, and it is from this point in time that her relationship with the parrots became a serious undertaking. Pam and the scientists would spend time painstakingly making their way through small plots, taking note of every plant growing there, and collecting samples of seeds that they would later examine with a microscope. Pam said that every day for three years she watched and followed the birds, keeping track of their movements. She no longer keeps a written log of individual birds (although she does map, photograph, and monitor nests), but she knows roughly the size and characteristics of the birds near her house, and she knows their favorite nesting and feeding spots. Park ranger programs to monitor the parrots have not achieved the same success or level of detail in their monitoring as Pam. Pam had been hoping to train some Olkola rangers at some of the nearby parks to work with the parrots, but for a variety of reasons this had not yet eventuated during the time I was in Cape York. Pam argues that getting to know the parrots cannot be done by working nine to five, five days a week. Over many years of early starts and through carefully paying attention, Pam has become attuned to the sounds, movements, and indicators of the parrots. She knows their daily and seasonal routes and maps where their nests are in a GPS database, complete with photographs and handwritten notes. You have to watch them and spend time with them for a long time to get to know them, Pam told me. Pam's knowledge is observational, embodied, and intuitive. Her ears and eyes are keenly attuned to signs in the landscape that indicate the presence of parrots—the grasses they like to feed on, the presence of other birds.

I accompanied Pam on many of her bird-watching tours. On one occasion, a couple that was caretaking a nearby national park came by to see the parrot. Pam piled us all into her car and took us to a good spot for seeing birds in the morning, down by the main road. Once we arrived and got out of the car, walking around softly, Pam walked away from the group. She would pause, listening, and then head in a slightly different direction. She knows the call of the parrot. "The young males sing out a lot, they're the easiest ones to hear," she told us.

Pam's manner with birders tended to remain consistent, whether they were Parks employees, dedicated bird enthusiasts, or wealthy tourists. She spoke quietly and didn't provide much direction. She would simply quietly point out the birds when she saw them, commenting on their sex and how mature they were likely to be, based on their coloring. Pam described how earlier in the morning, around 7:30 a.m., the parrots would spend time flitting around the trees by the road and preening the dew from their feathers. From there they would move along the fence line, eating grass seeds among the recently burned grass. Pam explained that she and her family would burn for the parrots so that they can find seed more easily and be a little safer from predators, burning each patch every two or three years. In the midafternoon, the parrots tended to stop by the bird feeder Pam had set up near her house. Pam explained that butcher birds and cats are the worst predators for the parrot, but chicken hawks are also a problem. Pam talked about the nesting habits of the birds, pointing out the kind of termite mounds they nest in. She said that during their nesting time, they lay one egg every day and a half, and if the air is particularly moist the termites will cover the nest hole with a thin layer of sand that the birds don't know how to dig through. In general, the bird-watchers Pam took on tours would be hushed, eager to spot a mature male and desperate to get the best photographs. The bird-watchers dressed in neutral tones. They carried large cameras with long lenses and binoculars on harnesses. There was always the distinct smell of sunscreen and insect repellent.

Success with spotting the parrot, though, and securing the sought-after photographs, relied on the bird-watchers behaving in particular ways. On one occasion, Pam took a group of doctors bird-watching at a small dam near her house that is frequented by bar-shouldered doves, finches, peewees, chicken hawks, wallabies, and golden-shouldered parrots. The leader of their group had visited once, a decade earlier, and was so taken with Pam and the parrots that he had arranged to bring a large group of friends and colleagues to take part in one of Pam's tours. The group was excited and chatty. Pam instructed them all to sit very still, as the parrots are far more disturbed by movement than by sound, but they moved around so much that a large flock of parrots flew off before anyone could photograph them or observe them with binoculars. Pam told the doctors that they were like "the biggest mob of kids," gently admonishing them and reminding them that they had to behave in a certain way if they wanted to see the parrots. These bird-watchers—with their fancy gear and on their expensive holiday—did not orient themselves in the proper way and so missed out on a close encounter with the parrot.

On multiple occasions, Pam told me that cattle grazing is the biggest threat to the parrots. She said that it used to be a less significant problem before they had fences. This is because at that time, the cattle could roam freely, but fences, along with the introduction of nutritional supplements, contributed to the cattle eating out paddocks. Fences directed and intensified their grazing in particular places as well as disrupting the route they may have taken in the past. If a fence appeared in the way of the route the cattle used to take, Pam explained to me, they won't look around for a different route, they will just remain where they are and eat out that paddock. Pam said that this overgrazing led to a decline in cockatoo grass—the main food source for the parrots.

The more substantial problem with overgrazing, though, lies in its relation to woody thickening of the savanna. When the grass is over-grazed, there is not enough fuel to have a hot fire when it comes time to burn, and melaleuca encroachment occurs. This results in a denser scrub growing around the termite mounds where parrots nest, which makes it harder for the parrots to see and evade predators. With more trees around, there are far more threats to the parrots. According to Pam, butcher birds and chicken hawks can sit in wait in trees above nests, and a parrot might not be able to see far enough beyond their nest to know whether or not it is a safe time to fly out.

Pam told me that once, a scientist had rigged up a surveillance camera by a nest and they could see a chicken hawk sitting terribly still, in wait, for hours. Pam said that often when she visits a nest, a butcher bird will be squawking above her, thinking that she's competing for food. On one occasion that she recalled, she knew a baby bird was going to fly soon and was keeping her eye on a butcher bird who, in turn, was keeping an eye on the nest. Pam wanted to see what the butcher bird would do. When the baby parrot poked its head out of the nest and attempted its first flight, the butcher bird attacked immediately, colliding with the chick in midair and knocking it to the ground. However, Pam had also started running for the butcher bird and successfully scared it away, scooping up the baby bird and returning it to the nest. "He probably just ate it the next day, though," Pam reflected.

The central tension in Pam's work with the parrots is the fact that cattle grazing has devastating impacts on parrot populations. Pam knows this. She openly discusses it with tourists, the visiting scientists, with me. Yet, her work with both species continues—the cattle on whom she and her family rely, and the parrots, who in some way have become entwined with Pam. The relationship is neither pure nor able to be neatly delineated, and it leads us to question how we consider who and what belongs in a region like Cape York. Cattle are, as Deborah Bird Rose put it, "four-legged soldiers in the army of conquest" (Rose 2004, 86). Pam herself is a settler-descended white woman who is implicated in the colo-nization of Cape York, and in the transformation of the landscape under the soil-compacting hooves of the cattle. In her own lifetime, the station that she lives on benefited directly from un- and underpaid Aboriginal labor. If we consider Cape York to be a blasted landscape, or at least a site of disturbance (Tsing 2015), we have to see Pam as a part of these histories and continuities of colonialism, violence, and environmental harms.

Pam's relationship to the parrots represents the kind of "constitutive non-coherence" that Shotwell talks about: "the fact that the world is made up of things that seem to hang together but require work to hold in place" (2016, 14). In her relationship with the parrots Pam has both proximity and intimacy—and here her work departs from common approaches to conservation. Pam clears trees to reduce the threat of predators for the parrots, and she uses the income from her bird-watching tours to buy seed with which she fills several homemade bird feeders. Pam exemplifies the difference between "caring about," that is, worrying for in an abstract and detached way, and "caring for," that is, doing the tedious maintenance work that relations of care require (Puig de la Bellacasa 2017). Her untidy and imperfect acts of care for the parrot are a shift away from the politics of purity that constrain formal conservation projects. Aware of her complicity but not paralyzed by it, Pam's work remains messy, entangled, and, ultimately, generative.

Pam's relationship to the parrots is emblematic of the complex practices of care that characterize the Cape York region. While implicated in various forms of harm toward the parrot by virtue of her engagement in the cattle industry, Pam labors to enact care for the parrot. To the best of her ability, she tries to allow for mutual flourishing. She is not concerned with the idea of protecting or creating a "wilderness," and despite her love for the parrot she is not suggesting that cattle should be removed from Cape York. Aware of the contradictions in her work, Pam's approach is pragmatic.

While Pam's relationship to the parrot is unique among her contemporaries, it is emblematic of the complicated forms of care enacted by the land managers whose stories populate this book. Like Pam, many of the cattle graziers, Aboriginal Traditional Owners, and park rangers of Cape York seek to cultivate what I have come to think of as workable landscapes. Workable landscapes in a settler-colony like Australia cannot be a pristine wilderness, made up of pure, natural, nativeness. These landscapes bear the scars of violent colonization, of agro-industry, of conservation regimes. People working on and with the land in Cape York live among inheritances—of knowledge, of wealth, of trauma, of ecological degradation. Their work is simultaneously shaped by these histories and their own imagined futures of the region. This is true whether these land managers are Aboriginal Traditional Owners, cattle graziers, or government-employed park rangers.

As this book has shown, although asymmetries run deep in the social fabric of Cape York, environmental knowledge, practices, forms of care, and even values are coconstituted in shifting intercultural and interspecies assemblages.

Cattle, having functioned as agents of colonization, continue to mediate human-environment relationships, for Aboriginal and settler-descended people alike. They reveal fissures and fractures in the relationship between established locals and the park rangers who are read as representatives of the government, as the face of the state's meddling in people's lives. Yet, in the relationship between (in particular) graziers and cattle, an ethic of care is evident *within and at the same time* as an instrumental human-animal relationship that, necessarily, involves forms of both violence and control. Cattle are understood by graziers and many Aboriginal Traditional Owners with historical and contemporary relationships to the grazing industry to contingently "belong" in Cape York in some sense, and much of the care that is enacted toward the land is done so, in part, to enable the flourishing of cattle in this landscape alongside other native species.

Within National Parks, though, cattle do not belong; they are considered "creatures out of place" (Scaramelli 2021) and are the target for removal and control, initiating hostility between Queensland Parks and graziers in particular, but also generating some tensions between Queensland Parks and Aboriginal Traditional Owners who would prefer to see cattle management carried out differently. The question whether running cattle and caring for land are mutually exclusive or not brings to the fore the different ways in which conservation organizations and people who live off the land in Cape York think about "good" land management. Bound and constrained by departmental requirements and legislative mandates, Queensland Parks' rangers seek to cultivate a park that is free of cattle, seeing their presence as detrimental and threatening to the natural and cultural values of the park. For graziers, and many Aboriginal Traditional Owners, though, there is both a desire and a freedom to engage in caring labor to create workable landscapes that support natural, cultural, and economic values simultaneously. In a "livable nature" (Scaramelli 2021), or a workable landscape, the presence of cattle does not indicate that land is being cared for poorly.

Pests and weeds provoke questions about how introduced species are categorized and subsequently controlled, revealing what kinds of workable landscapes are being sought and supported. In pest and weed control, violence intermingles with care, and some of the complexity involved in the muddy and quotidian work of caring for land becomes apparent. The relationship between care and complicity can be teased out in thinking about weed control, particularly inside National Parks. Aware that their vehicles and clothes transport seeds around the park, rangers nonetheless persist in their laboring to control the ever-expanding footprints of invasive plant species. They follow the guidance of senior management in focusing their efforts on particular weeds, aware that others will continue to proliferate. Rangers know that their work to control weeds is having little impact; they know that their small workforce is no match for the spread of these highly effective invasive plants across the vast landscape of the park. And yet, like other land managers in Cape York, they persist, and they make do.

Across the diversity of land managers in Cape York, controlling invasive plant and animal species is an act of contingent and complicated care. The violence enacted toward some species is done in order to create the conditions to allow other species, ecosystems, and ancestral spirits to endure. Yet, as the discussion on pig control demonstrates, only certain people's actions are construed as killing as a form of care in formal conservation settings; pig hunters, in contrast, are deemed unauthorized actors, exhibiting behavior that is uncaring, even as they are working to kill the same problematic pest species. The forms of care that emerge in invasive species control are the result of grappling with inherited landscapes, shaped by previous human efforts and nonhuman traces. While some invasive species are colonial remnants, or the aftereffects of particular economic projects, others have arrived in Cape York of their own volition.

In invasive species management, things coagulate and come together; inheritances, ongoing complicity, the desire to control and create particular kinds of landscapes, and forms of care rub up against each other.

Finally, the elements of fire and water, which everyone must work with and respond to, demonstrate how environmental knowledge and land management practices are coproduced interculturally, while also revealing how the same event or practice can be understood in multiple ways. In managing fire and dealing with water, Cape York residents are interacting with the impacts of climate change and with some of the strategies used to mitigate it. Cool burns are a land management technique and form of enacting care for Country that Aboriginal people in Cape York have used for millennia. For graziers and National Parks, these burning practices have a much shorter—although not insubstantial—history. Aboriginal people in Cape York see burning as part of a long cultural tradition that enacts care for the ancestral spirits that dwell in the landscape; by burning the land, their ongoing presence and connection to place is reaffirmed and their care for the "old people" there is made explicit. This kind of burning enacts care at multiple scales. Burning is care for ancestors, for the ongoing transmission of traditional knowledge and well-being of future generations, for the fire-adaptive and dependent species that make up some of these ecosystems, and—although rarely articulated in this way—for the planet as whole through the role that cool burns play in carbon abatement projects. Perhaps because of its local significance, talk about fire never really crept into talk about the climate. However, talk about the wet season and changes to weather patterns was where climate change became a point of discussion, and contention. While Aboriginal Traditional Owners engage strategically with the lexicon of climate change, synthesizing this language and these ideas into their own systems of knowing and observing the weather, graziers are resistant to the concept of climate change. They see climate change as bound up in a "green" ideology—one that, importantly, implicates them and their industry in environmental harms. In pushing back against the idea of climate change, though, and refusing to acknowledge their complicity in climate change, graziers still demonstrate proenvironmental values through their actions. They still enact tangible forms of care for the land, even if their rhetoric about the climate veers into the realm of uncaring.

Land management practices across pastoral lease, Aboriginal land, and National Parks are not discrete, separate bodies of knowledge or work. Instead, knowledge is shared and contingently coproduced. Much is held in common. Between the shifting assemblages of people and land tenure types, laboring on and with the land emerges as vital. This is the case even—and, perhaps especially so—where such laboring and care is noninnocent, compromised, and interested.

As I said at the start of this book, I am not interested in thinking about ways to do conservation better. The structures and logic of conservation leave too many ways of relating to other people, landscapes, and the more-than-human world out of the picture. The politics of purity that conservation regimes are shaped by fall flat in a place like Cape York, a place characterized by high and highly contested environmental, cultural, and economic values. Instead, in thinking through what it means to stick with care and complicity, through the framework of making do, it is possible to see how a diversity of people in Cape York labor to make livable not only their own lives but the lives of the nonhumans with which they share this place.

Throughout this book, I have attended to care as a labor practice. I am less concerned with the ways in which people care *about* animals, plants, ecosystems, and livelihoods, and more interested in the work they do to enact multiple forms of care toward these things. I have conceptualized caring labor quite broadly, allowing it to encompass actions that we may think of as uncaring—such as killing and culling, and even moments of interpersonal conflict—as well as the practices that we might expect indicate care for the land. Part of my contribution to the work on care has been to pay attention to the relationship between holding a position that may appear to be uncaring and the actual care labor and work that people do. The caring practices and labor of the people in this book often confound the categories through which they are trying to make sense of their place in the world. A contradiction is inherent here, but as Davé (2023) has pointed out, identifying where a contradiction is present only gets us so far. Perhaps what is more interesting, and productive, is to think about how the people who live with such contradictions don't get stuck. They are not caught up in the search for ideological or ethical purity. They can get on with the caring work that needs to be done—even if certain practices seem to undermine their stated position, even if the caring labor remains partial, unfinished, and less than satisfactory.

I do not conceive of care and self-interest as mutually exclusive. Indeed, the spaces in which care, self-interest, and complicity in environmental harms intermingle are those that I have chosen to attend to in detail. In the still-remote and economically marginal region of Cape York, care for land has to happen alongside a consideration of livelihoods. If people cannot afford to live in the region, they will not be there to engage in the caring practices that the ecosystems, nonhuman species, and ancestral spirits need. Puig de la Bellacasa asks "for what worlds is care being done for?" (2017, 64–65), and here, in this place, I have argued that care is being done to bring into being sustainable futures: livable worlds and workable landscapes. For the people who live in Cape York, these futures have to be both economically viable and environmentally sound. Cape York is a place in which people simultaneously feel the impacts of perceived government overreach and neglect; it is a place where people—graziers, park rangers, and Traditional Owners alike—value their freedom and autonomy. People labor to create workable landscapes, where they can make their lives and livelihoods while *at the same time* enact care for the land, for ecosystems, and for the nonhumans with which they share this place. In making workable landscapes, compromises are made, and contradictions are present. I have followed the lead of my interlocutors in staying with care, complicity, and the space between them—a space that I think of as making do—to think about how we can envision human relationships to the more-than-human world outside of the constraints of conservation. Impure, imperfect, messy, but enduring.

Like Pam who loves her cattle and her parrots at the same time, Cape York land managers labor, persistently, to make do amid contradictions and constraints.

Bushfire smoke, Rinyirru National Park, 2020.

NOTES

Introduction

1. Throughout this book I use pseudonyms to refer to people and cattle stations to protect the identities of my interlocutors. I do, however, use the actual names of national parks and language groups.

2. "Scrub" is a word colloquially used in Australia, interchangeable with "bush" or "brush," that refers to uncultivated land.

3. However, it is important to note that the positive impacts of de-stocking may take decades rather than years to materialize, as increased groundcover alone is not sufficient to reduce sediment runoff; rather, deep-rooted species need to be allowed to establish (Bartley et al. 2014).

4. Both of the national parks described in this book are jointly managed, although my focused analysis of these joint management arrangements appears elsewhere (see Reardon-Smith 2024; 2025).

Chapter One

1. Trap paddocks are fenced sections of land with a contraption called a "spear trap" at the entrance. This spear trap is made of angled pieces of wood or metal, positioned in such a way that cattle can enter the paddock easily but cannot exit. Trap paddocks are normally constructed around a waterpoint or a salt lick (nutritional supplement) in order to lure cattle in. The use of trap paddocks means that graziers are able to locate and move their stock with smaller workforces than in the past as the cattle tend to end up in these centralized locations.

2. A "ringer" is a gender-neutral term for a cowboy/cowgirl or jackaroo/jillaroo. It is the vernacular term most people in Cape York and other regions in Australia frequently use to describe someone who does stock work.

3. Throughout the nineteenth century, Australia experienced many "gold rushes" that brought an influx of immigrants to the continent, hoping to make their fortunes mining gold. Gold rushes—like the one that occurred in the Palmer River region of southern Cape York—were often short-lived, necessitating the rapid construction (and often subsequent abandonment) of makeshift towns.

4. The Aboriginals Protection and Restriction of the Sale of Opium Act 1897 tightly controlled multiple aspects of Aboriginal peoples' lives in Queensland, ensuring that Aboriginal people were either on reserves or in employment (May 1994).

5. The forebears of many of the Lama Lama people that I met were part of the community who were forcibly removed to a mission at Bamaga and later Injinoo for ease of government administration, where people mostly remained for nearly three decades before reestablishing a community at Port Yintjingga.

6. See, for instance, Merlan (1998), Ottosson (2012), and Redmond (2005).

7. In his work with Cape York Wik people, Peter Sutton (1978) noted that the word denoting ownership, *kooepanha*, actually means to "look after," "wait on" or "guard." He described how people demonstrate their ownership, or more accurately custodianship, through fulfilling obligations to the land. Writing around the same time, John Von Sturmer (1978) also detailed Wik people caring for country through following protocols that help to ensure the sustainability of exploitable resources. In southeastern Cape York, Anderson (1985) suggested that Kuku-Yalanji people perceived this situation of mutual obligation and custodianship as fostering an affective relationship that transcended their economic relationships with the land. Likewise, David Martin (1993) suggested that Wik people see their landscape as socialized through relationships formed by consistent activities like hunting, fishing, and camping.

Chapter Two

1. The *Courier Mail* is a widely read newspaper in Queensland, owned by News Corp Australia.

2. This is particularly the case for White Australia. Brumbies are considered by many to be an iconic species, evocative in some way of the (White) Australian spirit and appearing in books and films, notably the Banjo Patterson poem "The Man From Snowy River" and the book and film *The Silver Brumby*.

3. Border Force is an Australian federal law enforcement agency, responsible for the policing of Australia's borders. Their presence at the grazing forum is indicative of Cape York's proximity to the Torres Strait and Australia's northern neighbors, and Cape York's status as a kind of frontier.

Chapter Three

1. "Lick" is the term graziers use to describe nutritional supplements fed to their cattle. See chapter one for a discussion of lick.

2. Cape York Natural Resource Management is a not-for-profit organization that helps landholders carry out sustainability-focused projects and access government grants.

3. As the local government authority, the Cook Shire Council provides land-holders in the southern part of Cape York with funding to carry out weed management projects that align with federal and state government priorities.

4. Biosecurity Queensland is a government service separate from QPWS, although some of their work is carried out on national parks.

5. While buffel grass is a declared weed and has been described as a threat to biodiversity in parts of South Australia, Western Australia, the Northern Territory and central Queensland, it is not a species of concern in Cape York (CRC for Australian Weed Management 2008).

Chapter Five

1. Approximately 1.2 miles.

2. In the management of the Gran Sabana in Venezuela, for example, Sletto (2008) has demonstrated how indigenous burning practices have been historically denigrated by state agencies as irrational and destructive. Such a position is based on an imagined past of the Gran Sabana as a forest that has been negatively affected by indigenous burning and is at risk of becoming a desert in the future (Sletto 2011; 2008). Throughout the nineteenth and twentieth centuries, fire regimes were similarly suppressed in various parts of Africa, including Madagascar (Kull 2002), Zambia (Eriksen 2007), Mali (Laris 2002), and Mozambique (Shaffer 2010). While Shaffer (2010) speaks to the sidelining of Indigenous peoples' traditional ecological knowledge in state-led fire management, Kull (2002), Eriksen (2007), and Laris (2002) variously describe how state conservation agencies built on earlier colonial policies to prohibit and criminalize the use of fire, positioning indigenous fire regimes as disruptive, dangerous, and antithetical to conservation. Since the late aughts, conservation discourse, strategies, and policies in various sub-Saharan African and South American states have begun to shift toward an acceptance of fire management rather than suppression (Eloy et al. 2019; Moura et al. 2019; Schmidt and Eloy 2020). However, it is important to note that there remains a diversity of opinion among conservation and land management actors about the benefits or risks of fire management (Rodríguez et al. 2018; Eloy et al. 2019, 16) and, furthermore, many fire management projects are reliant on international development funding (Moura et al. 2019, 602).

REFERENCES

ABC News. 2019. "Tropical Cyclone Trevor Continues across Far North Queensland as Category One System." *ABC News*, March 20, 2019. Accessed January 14, 2020. www.abc.net.au/news/2019-03-20/cyclone-trevor-tracks-across-cape-york-peninsula-as-category-two/10916524.

Adams, W., and M. Mulligan. 2002. Introduction to *Decolonizing Nature: Strategies for Conservation in a Post-Colonial Era*, edited by W. Adams and M. Mulligan, 1–15. Taylor and Francis Group.

Addison, C. 2022. "1080." In *An Anthropogenic Table of Elements: Experiments in the Fundamental*, edited by T. Neale, C. Addison and T. Phan, Technoscience and Society, 22–33. University of Toronto Press.

Altman, Jon. 2012. *People on Country: Vital Landscapes/Indigenous Futures.* Federation Press.

Anderson, J. C. 1985. "The Political and Economic Basis of Kuku-Yulanji Social History." PhD diss., School of Social Science, University of Queensland.

———. 1989. "Aborigines and Conservation: The Daintree-Bloomfield Road." *Australian Journal of Social Issues* 24, no. 3: 214–27.

Anderson, Virginia DeJohn. 2004. *Creatures of Empire: How Domestic Animals Transformed Early America.* Oxford University Press.

Arcoverde, Gabriela B., Alan N. Anderson, and Samantha A. Setterfield. 2017. "Is Livestock Grazing Compatible with Biodiversity Conservation? Impacts on Savanna Ant Communities in the Australian Seasonal Tropics." *Biodiversity and Conservation* 26: 883–97. https://doi.org/10.1007/s10531-016-1277-5.

Arnaquq-Baril, A. 2016. Angry Inuk. ONF/NFB, Canada.

Arregui, A. G. 2023. "Reversible Pigs: An Infraspecies Ethnography of Wild

Boars in Barcelona." *American Ethnologist*: 1–14. https://doi.org/10.1111/amet.13114.

Atchison, J., and L. Head. 2013. "Eradicating Bodies in Invasive Plant Management." *Environment and Planning D: Society and Space* 31: 951–68. https://doi.org/10.1068/d17712.

Australian Government. 2012. *A World Heritage Nomination for Cape York Peninsula*. Department of Sustainability, Environment, Water, Population, and Communities, ACT, Canberra. www.dcceew.gov.au/sites/default/files/documents/cape-york-nomination.pdf.

Babidge, Sally. 2019. "Sustaining Ignorance: The Uncertainties of Groundwater and Its Extraction in the Salar de Atacama, Northern Chile." *Journal of the Royal Anthropological Institute* 25, no. 1: 83–102. https://doi.org/10.1111/1467-9655.12965.

Bach, T. M., and B. M. H. Larson. 2017. "Speaking about Weeds: Indigenous Elders' Metaphors for Invasive Species and Their Management." *Environmental Values* 26, no. 5: 561–81.

Baker, R. M. 1999. *Land Is life: From Bush to Town; The Story of the Yanyuwa People*. Allen and Unwin.

Balée, W. 2013. *Cultural Forests of the Amazon: A Historical Ecology of People and Their Landscapes*. University of Alabama Press.

Bankoff, Greg. 2003. "Constructing Vulnerability: The Historical, Natural, and Social Generation of Flooding in Metropolitan Manila." *Disasters* 27, no. 3: 224–38.

Bartley, Rebecca, Jeff P. Corfield, Aaron A. Hawdon, Anne E. Kinsey-Henderson, Brett N. Abbott, Scott N. Wilkinson, and Rex J. Keen. 2014. "Can Changes to Pasture Management Reduce Runoff and Sediment Loss to the Great Barrier Reef? The Results of a 10-Year Study in the Burdekin Catchment, Australia." *Rangeland Journal* 36, no. 1: 67–84. https://doi.org/10.1071/RJ13013.

Beldo, Les. 2019. *Contesting Leviathan: Activists, Hunters, and State Power in the Makah Whaling Conflict*. University of Chicago Press.

Bengsen, A. J., P. West, and C. R. Krull. 2018. "Feral Pigs in Australia and New Zealand: Range, Trend, Management, and Impacts of an Invasive Species." In *Ecology, Conservation, and Management of Wild Pigs and Peccaries*, edited by M. Melletti and E. Meijaard, 325–38. Cambridge University Press.

Bennet, D., and C. Sheehan. 2021. "Koowarta, John Pampeya (1940–1991)." *Australian Dictionary of Biography* 19. https://adb.anu.edu.au/biography/koowarta-john-pampeya-14856.

Berkes, F. 1993. "Traditional Ecological Knowledge in Perspective." In *Traditional Ecological Knowledge: Concepts and Cases*, edited by J. T. Inglis, 1–10. International Development Research Centre.

Bessire, Lucas. 2021. *Running Out: In Search of Water on the High Plains*. Princeton University Press.

Blanchette, A. 2020. *Porkopolis: American Animality, Standardized Life, and the Factory Farm*. Duke University Press.

Blok, A. 2011. "Clash of the Eco-sciences: Carbon Marketization, Environmental NGOs, and Performativity as Politics." *Economy and Society* 40, no. 3: 451–76.

Bocci, P. 2017. "Tangles of Care: Killing Goats to Save Tortoises on the Galapagos Islands." *Cultural Anthropology* 32, no. 3: 424–49. https://doi.org/10.14506/ca32.3.08.

Bonifacio, V. 2023. "Of Feral and Obedient Cows: Colonization as Domestication in the Paraguayan Chaco." *Cultural Anthropology* 38, no. 1: 8–35. https://doi.org/10.14506/ca38.1.02.

Bowman, D. M. J. S. 1998. "Tansley Review No. 101: The Impact of Aboriginal Landscape Burning on the Australian Biota." *New Phytologist* 140, no. 3: 385–410.

Bradley, J. 1995. "Fire: Emotion and Politics: A Yanyuwa Case Study." In *Country in Flames: Proceedings of the 1994 Symposium on Biodiversity and Fire in North Australia*, edited by D. B. Rose, Biodiversity Series, 25–32. Canberra Department of Environment, Sport and Territories.

Braun, Bruce. 2002. *The Intemperate Rainforest: Nature, Culture, and Power on Canada's West Coast*. University of Minnesota Press.

Braverman, Irus. 2018. *Coral Whisperers: Scientists on the Brink*. Critical Environments: Nature, Science, and Politics. University of California Press.

Brice, J. 2014. "Attending to Grape Vines: Perceptual Practices, Planty Agencies, and Multiple Temporalities in Australian Viticulture." *Social and Cultural Geography* 15, no. 8: 942–65. https://doi.org/10.1080/14649365.2014.883637.

Brodie, J. E., F. J. Kroon, B. Schaffelke, E. C. Wolanski, S. E. Lewis, M. J. Devlin, I. C. Bohnet, Z. T. Bainbridge, J. Waterhouse, and A. M. Davis. 2012. "Terrestrial Pollutant Runoff to the Great Barrier Reef: An Update of Issues, Priorities, and Management Responses." *Marine Pollution Bulletin* 65, nos. 4–9: 81–100. https://doi.org/https://doi.org/10.1016/j.marpolbul.2011.12.012.

Bubandt, N., and A. Tsing. 2018. "Feral Dynamics of Post-Industrial Ruin: An Introduction." *Journal of Ethnobiology* 38, no. 1: 1–7.

Carroll, Clint. 2015. *Roots of Our Renewal: Ethnobotany and Cherokee Environmental Governance*. First Peoples: New Directions in Indigenous Studies. University of Minnesota Press.

Caruana, P. 2012. "Cape Traditional Owners Win 38-Year Battle." *Sydney Morning Herald*, May 22, 2012. www.smh.com.au/national/cape-traditional-owners-win-38-year-battle-20120522-1z1t6.html.

Casagrande, D. G., H. McIlvaine-Newsad, and E. C. Jones. 2015. "Social Networks of Help-Seeking in Different Types of Disaster Responses to the 2008 Mississippi River Floods." *Human Organization* 74, no. 4: 351–61.

Cattelino, J. 2017. "Loving the Native: Invasive Species and the Cultural Politics of Flourishing." In *The Routledge Companion to the Environmental Humanities*, edited by U. K. Heise, J. Christensen, and M. Niemann. Routledge.

Charnely, Susan, Thomas E. Sheridan, and Gary P. Nabhan. 2014. Introduction to *Stitching the West Back Together: Conservation of Working Landscapes,* edited by Susan Charnley, Thomas E. Sheridan, and Gary P. Nabhan, xiii–xxi. University of Chicago Press.

Chase, A. K. 1980. "Which Way Now? Tradition, Continuity, and Change in a North Queensland Aboriginal Community." PhD diss., School of Social Science, University of Queensland.

Chester, G. 2010. *Cape York World Heritage: Discussion Paper Prepared for Cape York Sustainable Futures*. EcoSustainAbility Pty.

Coghlan, Lea. 2017. "Graziers Raise Concerns with Cattle Culling in National Parks." *North Queensland Register*, 2017. Accessed June 23, 2023. www.northqueenslandregister.com.au/story/4664341/cattle-culling-concerns/.

Cole, N. 2004. "Battle Camp to Boralga: A Local Study of Colonial War on Cape York Peninsula, 1873–1894." *Aboriginal History* 28: 156–89.

Commonwealth of Australia. 2020. *Australia's Sixth National Report to the Convention on Biological Diversity, 2014–2018*. Department of Agriculture, Water and the Environment. ACT, Canberra. www.dcceew.gov.au/sites/default/files/documents/sixth-national-report-convention-biological-diversity.pdf.

Connor, L. H. 2016. *Climate Change and Anthropos: Planet, People and Places*. Routledge.

Connor, L. H., and N. Higginbotham. 2013. "'Natural Cycles' in Lay Understandings of Climate Change." *Global Environmental Change* 23, no. 6: 1852–61.

Convention on Biological Diversity. June 5, 1992. 1760 U.N.T.S. 69, ATS 32.

Cousins, D. V., and J. L. Roberts. 2001. "Australia's Campaign to Eradicate Bovine Tuberculosis: The Battle for Freedom and Beyond." *Tuberculosis* 81, no. 1/2: 5–15. https://doi.org/10.1054/tube.2000.0261.

Crate, S. A. 2008. "Gone the Bull of Winter? Grappling with the Cultural Implications of and Anthropology's Role(s) in Global Climate Change." *Current Anthropology* 49, no. 4: 569–95.

Crate, S. A., and M. Nuttall. 2016. "Introduction: Anthropology and Climate Change." In *Anthropology and Climate Change: From Encounters to Actions*, edited by S. A. Crate and M. Nuttall, 11–34. Routledge.

CRC for Australian Weed Management. 2008. Weed Management Guide: Buffel Grass (*Cenchrus ciliaris*). www.aabr.org.au/images/stories/resources/ManagementGuides/WeedGuides/wmg_buffelGrass.pdf.

Cronon, W. 1996. "The Trouble with Wilderness: Or, Getting Back to the Wrong Nature." *Environmental History* 1, no. 1: 7–28.

Crowther, M. S., M. Fillios, N. Colman, and M. Letnic. 2014. "An Updated Description of the Australian Dingo (*Canis dingo* Meyer, 1793)." *Journal of Zoology* 293: 192–203.

Cruikshank, J. 2005. *Do Glaciers Listen? Local Knowledge, Colonial Encounters, and Social Imagination*. University of British Columbia Press.

Dalsgaard, S. 2013. "The Commensurability of Carbon: Making Value and Money of Climate Change." *HAU: Journal of Ethnographic Theory* 3, no. 1: 80–98.

Daniels, M. J., and L. Corbett. 2003. "Redefining Introgressed Mammals: When Is a Wildcat a Wild Cat and a Dingo a Wild Dog?" *Wildlife Research* 30: 213–18.

Davé, Naisargi N. 2017. "Witness: Humans, Animals, and the Politics of Becoming." In *Unfinished: The Anthropology of Becoming*, edited by J. Biehl and P. Locke, 151–69. Duke University Press.

———. 2023. *Indifference: On the Praxis of Interspecies Being*. Duke University Press.

Davis, J. 2003. "Indigenous Land Management." In *Australia Burning: Fire, Ecology, Policy, and Management Issues*, edited by G. Cary, D. Lidenmayer, and S. Dovers, 219–23. CSIRO Publishing.

De Rijke, K. 2012. "Water, Place and Community: An Ethnography of Environmental Engagement, Emplaced Identity and the Traveston Crossing Dam Dispute in Queensland, Australia." PhD thesis, University of Queensland.

Department of Agriculture and Fisheries. 2017. "Toxin 1080: A Guide to Safe and Responsible Use of Sodium Fluoroacetate in Queensland." State of Queensland.

Department of Environment and Science. 2019. "Climate Change in the Cape York Region." State of Queensland.

———. 2021a. "Oyala Thumotang National Park (CYPAL): Nature, Culture, and History." Queensland Government. Accessed February 8. https://parks.des.qld.gov.au/parks/oyala-thumotang/about/culture.

———. 2021b. *Queensland: State of the Environment 2020 Summary*. State of Queensland. www.stateoftheenvironment.des.qld.gov.au/_media/documents/Queensland-State-of-the-Environment-2020-Summary.pdf.

Descola, P. 2013. *Beyond Nature and Culture*. University of Chicago Press.

Descola, P., and G. Pálsson. 1996. *Nature and Society: Anthropological Perspectives*. Routledge.

Diekmann, Lucy, Lee Panich, and Chuck Striplen. 2007. "Native American Management and the Legacy of Working Landscapes in California: Western Landscapes Were Working Long before Europeans Arrived." *Rangelands* 29, no. 3: 46–50.

Dombrowski, K. 2001. *Against Culture: Development, Politics, and Religion in Indian Alaska*. University of Nebraska Press.

Dominy, M. 1997. "The Alpine Landscape in Australian Mythologies of Ecology and Nation." In *Knowing Your Place: Rural Identity and Cultural Hierarchy*, edited by B. Ching and G. W. Creed, 237–66. Routledge.

———. 2001. *Calling the Station Home: Place and Identity in New Zealand's High Country*. Rowman and Littlefield.

Dominy, M. D. 2003. "Hearing Grass, Thinking Grass: Postcolonialism and Ecology in Aotearoa-New Zealand." In *Disputed Territories: Land, Culture and Identity in Settler Societies*, edited by D. Trigger and G. Griffiths, 53–80. Hong Kong University Press.

Doody, B. J., H. C. Perkins, J. J. Sullivan, C. Meurk, and G. H. Stewart. 2014. "Performing Weeds: Gardening Plant Agencies and Urban Plant Conserva-

tion." *Geoforum* 56: 124–36. https://doi.org/10.1016/j.geoforum.2014.07
.001.

Doolittle, Amity A. 2005. *Property and Politics in Sabah, Malaysia: Native
Struggles over Land Rights.* Culture, Place and Nature. University of Washington Press.

Einarsson, N. 1993. "All Animals Are Equal but Some Are Cetaceans: Conservation and Culture Conflict." In *Environmentalism: The View from Anthropology,* edited by K. Milton, 73–84. Routledge.

Eloy, L., B. A. Bilbao, J. Mistry, and I. B. Schmidt. 2019. "From Fire Suppression to Fire Management: Advances and Resistances to Changes in Fire
Policy in the Savannas of Brazil and Venezuela." *Geographical Journal* 185:
10–22.

Eriksen, C. 2007. "Why Do They Burn the 'Bush'? Fire, Rural Livelihoods, and
Conservation in Zambia." *Geographical Journal* 173, no. 3: 242–56.

Escobar, A. 2008. *Territories of Differences: Place, Movements, Life, Redes.*
Duke University Press.

Farley, Simon. 2023. "Mateship with Brumbies: Horses, Defiance, and Indigeneity in the Australian Alps." *Journal of Australian Studies* 47, no. 2: 256–72. https://doi.org/10.1080/14443058.2022.2142835.

Ficek, Rosa E. 2019. "Cattle, Capital, Colonization: Tracking Creatures of the
Anthropocene In and Out of Human Projects." *Current Anthropology* 60,
no. 20: S260–S271. https://doi.org/10.1086/702788.

Fischer, John Ryan. 2015. *Cattle Colonialism: An Environmental History of
the Conquest of California and Hawai'i.* University of North Carolina Press.

Førde, A., and T. Magnussen. 2015. "Invaded by Weeds: Contested Landscape
Stories." *Geografiska Annaler: Series B, Human Geography* 97, no. 2: 183–93.

Four Corners. 2011. "A Bloody Business." ABC News, Australia.

Galvin, S. S. 2018. "Interspecies Relations and Agrarian Worlds." *Annual Review
of Anthropology* 47: 233–49.

Gammage, B. 2012. *The Biggest Estate on Earth: How Aborigines Made Australia.* Allen and Unwin.

Garbutt, R. G. 2011. *The Locals: Identity, Place, and Belonging in Australia
and Beyond.* Peter Lang.

Geiger, Dominic. 2016a. "Call to Halt Planned Cattle Cull on Cape York."
Cairns Post, October 11. Accessed June 22, 2023. www.cairnspost.com.au/
news/cairns/call-to-halt-planned-cattle-cull-on-cape-york/news-story/
5671220c20f2b8cde904a5cf45127cef.

———. 2016b. "Cattle Cull to Start on Cape York Despite Animal Welfare Concerns." *Cairns Post,* October 10, 2016. Accessed June 23. www.cairnspost
.com.au/news/cairns/cattle-cull-to-start-on-cape-york-despite-animal
-welfare-concerns/news-story/7e5c2567be668d2208b17ce4af451c02.

Geschiere, P. 2009. *The Perils of Belonging: Autochthony, Citizenship, and
Exclusion in Africa and Europe.* University of Chicago Press.

Gibbs, L., J. Atchison, and I. Macfarlane. 2015. "Camel Country: Assemblage,

Belonging, and Scale in Invasive Species Geographies." *Geoforum* 58: 56–67. https://doi.org/10.1016/j.geoforum.2014.10.013.

Gibson, J. W. 2019. "Automating Agriculture: Precision Technologies, Agbots, and the Fourth Industrial Revolution." In *In Defense of Farmers: The Future of Agriculture in the Shadow of Corporate Power*, edited by J. W. Gibson and S. E. Alexander, 135–73. University of Nebraska Press.

Gill, N. 1997. "Pastoralism, a Contested Domain." In *Tracking Knowledge in North Australian Landscapes: Studies in Indigenous and Settler Ecological Knowledge Systems*, edited by D. B. Rose and A. Clarke, 50–67. North Australian Research Unit, Australian National University.

———. 2005. "Transcending Nostalgia: Pastoralist Memory and Staking a Claim in the Land." In *Dislocating the Frontier: Essaying the Mystique of the Outback*, edited by D. B. Rose and R. Davis, 67–84. Australian National University Press.

———. 2014. "Making Country Good: Stewardship and Environmental Change in Central Australian Pastoral Culture." *Transactions of the Institute of British Geographers* 39, no. 2: 265–77.

Gill, N., and A. Paterson. 2007. "A Work in Progress: Aboriginal People and Pastoral Cultural Heritage in Australia." In *Geographies of Australian Heritages: Loving a Sunburnt Country?*, edited by B. J. Shaw and R. Jones, 113–31. Taylor and Francis.

Govindrajan, R. 2018. *Animal Intimacies: Interspecies Relatedness in India's Central Himalayas*. University of Chicago Press.

Graham, Mary. 1999. "Some Thoughts about the Philosophical Underpinnings of Aboriginal Worldviews." *Worldviews: Environment, Culture, Religion* 3: 105–18.

Grandia, Liza. 2012. *Enclosed: Conservation, Cattle, and Commerce among the Q'eqchi' Maya Lowlanders*, edited by K. Sivaramakrishnan. Culture, Place, and Nature. University of Washington Press.

Gray, J. 1999. "Open Spaces and Dwelling Places: Being at Home on Hill Farms in the Scottish Borders." *American Ethnologist* 26, no. 2.

Hagis, E., and J. Gillespie. 2021. "Kosciuszko National Park, Brumbies, Law, and Ecological Justice." *Australian Geographer* 52, no. 3: 225–41. https://doi.org/10.1080/00049182.2021.1928359.

Hamilton, Sarah R. 2018. *Cultivating Nature: The Conservation of a Valencian Working Landscape*. University of Washington Press.

Haraway, D. 2008. *When Species Meet*. University of Minnesota Press.

Hastrup, K. 2012. "Anticipating Nature: The Productive Uncertainty of Climate Models." In *The Social Life of Climate Change Models: Anticipating Nature*, edited by K. Hastrup and M. Skrydstrupp, 1–29. Routledge.

———. 2014. Introduction to *Living with Environmental Change: Waterworlds*, edited by K. Hastrup and C. Robow, 20–27. London: Routledge.

———. 2016. "Building Foundations of Anthropology and Climate Change." In *Anthropology and Climate Change: From Encounters to Action*, edited by S. A. Crate and M. Nuttall, 35–57. Routledge.

Hayden, C. 2003. *When Nature Goes Public: The Making and Unmaking of Bioprospecting in Mexico.* Princeton University Press.

Haynes, C. 2009. "Defined by Contradiction: The Social Construction of Joint Management in Kakadu National Park." PhD thesis, School for Social and Policy Research, Charles Darwin University.

Head, L. 1994. "Landscapes Socialised by Fire: Post-Contact Changes in Aboriginal Fire Use in Northern Australia, and Implications for Prehistory." *Archaeology in Oceania* 29, no. 3: 172–81.

———. 2000. *Second Nature: The History and Implications of Australia as Aboriginal Landscape.* Syracuse University Press.

———. 2012. "Decentring 1788: Beyond Biotic Nativeness." *Geographical Research* 50: 166–78.

———. 2016. *Hope and Grief in the Anthropocene: Re-Conceptualising Human-Nature Relations.* Routledge.

Head, Lesley. 2017. "The Social Dimensions of Invasive Plants." *Nature Plants* 3, no. 6: 1–7.

Head, Lesley, Jennifer Atchison, and Catherine Phillips. 2015. "The Distinctive Capacities of Plants: Re-Thinking Difference via Invasive Species." *Transactions of the Institute of British Geographers* 40: 399–413. https://doi.org/10.1111/tran.12077.

Helmreich, S. 2005. "How Scientists Think; about 'Natives,' for Example: A Problem of Taxonomy among Biologists of Alien Species in Hawaii." *Journal of the Royal Anthropological Institute* 11, no. 1: 107–28.

Higginbotham, N., L. H. Connor, and F. Baker. 2014. "Subregional Differences in Australian Climate Risk Perceptions: Coastal versus Agricultural Areas of the Hunter Valley, NSW." *Regional Environmental Change* 14: 699–712.

Hinkson, M., and B. Smith. 2005. "Introduction: Conceptual Moves towards an Intercultural Analysis." *Oceania* 75, no. 3: 157–66.

Hitchcock, P., M. Kennard, B. Leaver, B. Mackey, P. Stanton, P. Valentine, E. Vanderduys, B. Wannan, W. Willmott, and J. Woinarski. 2013. *The Natural Attributes for World Heritage Nomination of Cape York Peninsula, Australia.* Independent Scientific Expert Panel for the Department of Sustainability, Environment, Water, Population, and Communities. www.dcceew.gov.au/sites/default/files/env/resources/5ab50983-6bb4-4d87-8298-f1bcf1ab652a/files/sciencepanelreport.pdf.

Hobart, Hiʻilei Julia Kawehipuaakahaopulani, and Tamara Kneese. 2020. "Radical Care: Survival Strategies for Uncertain Times." *Social Text* 38, no. 1: 1–16. https://doi.org/10.1215/01642472-7971067.

Hoffman, S. M., and A. Oliver-Smith. 1999. "Anthropology and the Angry Earth: An Overview." In *The Angry Earth: Disaster in Anthropological Perspective*, edited by A. Oliver-Smith and S. Hoffman, 1–16. Routledge.

Holmes, J. 2011a. "Contesting the Future of Cape York Peninsula." *Australian Geographer* 42, no. 1: 53–68.

———. 2011b. "Land Tenures as Policy Instruments: Transitions on Cape York Peninsula." *Geographical Research* 49, no. 2: 217–33.

Howitt, R., and S. Suchet-Pearson. 2006. "Rethinking the Building Blocks: Ontological Pluralism and the Idea of 'Management.'" *Geografiska Annaler: Series B, Human Geography* 88, no. 3: 323–35.

Huntsinger, Lynn, and Nathan F. Sayre. 2007. "Introduction: The Working Landscapes Special Issue." *Rangelands* 29, no. 3: 3–4.

IUCN. 2022. "IUCN Red List: Background & History." International Union for Conservation of Nature and Natural Resources. www.iucnredlist.org/about/background-history.

Jackson, S. 2006. "Sons of Which Soil? The Language and Politics of Autochthony in Eastern D.R. Congo." *African Studies Review* 49, no. 2: 95–123.

Jorgensen, D. 2016. "The Garden and Beyond: The Dry Season, the Ok Tedi Shutdown, and the Footprint of the 2015 El Niño Drought." *Oceania* 86, no. 1: 25–39.

Keil, P. G. 2021. "Rank Atmospheres: The More-Than-Human Scentspace and Aesthetic of a Pigdogging Hunt." *Australian Journal of Anthropology* 32: 96–113. https://doi.org/10.1111/taja.12382.

Kingsley, J. Y., M. Townsend, R. Phillips, and D. Aldous. 2009. "'If the Land Is Healthy . . . It Makes the People Healthy': The Relationship between Caring for Country and Health for the Yorta Yorta Nation, Boonwurrung and Bangerang Tribes." *Health and Place* 15: 291–99.

Kirner, Kimberly D. 2015. "The Cultural Heritage of Family Ranches." *Rangelands* 37, no. 2: 85–89. https://doi.org/10.1016/j.rala.2015.01.007.

Koci, Jack, Roy C. Sidle, Anne E. Kinsey-Henderson, Rebecca Bartley, Scott N. Wilkinson, Aaron A. Hawdon, Ben Jarihani, Christian H. Roth, and Luke Hogarth. 2020. "Effect of Reduced Grazing Pressure on Sediment and Nutrient Yields in Savanna Rangeland Streams Draining to the Great Barrier Reef." *Journal of Hydrology* 582 (124520). https://doi.org/https://doi.org/10.1016/j.jhydrol.2019.124520.

Koowarta v. Bjelke-Petersen. 1982. High Court of Australia.

Krause, F. 2013. "Seasons as Rhythms on the Kemi River in Finnish Lapland." *Ethnos* 78, no. 1: 23–46.

Kull, C. A. 2002. "Madagascar Aflame: Landscape Burning as Peasant Protest, Resistance, or a Resource Management Tool?" *Political Geography* 21: 927–53.

Kull, C. A., and H. Rangan. 2008. "Acacia Exchanges: Wattles, Thorn Trees, and the Study of Plant Movements." *Geoforum* 39: 1258–72. https://doi.org/10.1016/j.geoforum.2007.09.009.

Lamond, Julieanne. 2023. "Dead Horse Gap: Intergenerational Justice and the Culling of Horses in the Australian Alps." *Australian Humanities Review* 71: 87–95. https://doi.org/10.56449/14222876.

Langton, M. 1998. *Burning Questions: Emerging Environmental Issues for Indigenous Peoples in Northern Australia.* Centre for Indigenous Natural and Cultural Resource Management, Northern Territory University.

———. 2002. "The 'Wild,' the Market, and the Native: Indigenous People Face New Forms of Global Colonisation." In *Decolonizing Nature: Strategies for*

Conservation in a Post-Colonial Era, edited by W. Adams and M. Mulligan, 79–107. Taylor and Francis Group.

Laris, P. 2002. "Burning the Seasonal Mosaic: Preventative Burning Strategies in the Wooded Savanna of Southern Mali." *Human Ecology* 30, no. 2: 155–86.

Lea, T. 2012. "Contemporary Anthropologies of Indigenous Australia." *Annual Review of Anthropology* 41: 187–202.

Lepselter, Susan. 2016. *The Resonance of Unseen Things: Poetics, Power, Captivity, and UFOs in the American Uncanny*. University of Michigan Press.

Lévi-Strauss, Claude. 1962. *Totemism*. Translated by Rodney Needham. Merlin Press.

Liboiron, Max. 2021. *Pollution Is Colonialism*. Duke University Press.

Lien, M. E., H. A. Swanson, and G. Ween. 2018. "Naming the Beast—Exploring the Otherwise." In *Domestication Gone Wild: Politics and Practices of Multispecies Relations*, edited by M. E. Lien, H. A. Swanson, and G. Ween, 1–30. Duke University Press.

Lin, T.-C., J. A. Hogan, and C-T. Chang. 2020. "Tropical Cyclone Ecology: A Scale-Link Perspective." *Trends in Ecology and Evolution* 35, no. 7: 594–604.

Loos, N. 1982. *Invasion and Resistance: Aboriginal-European Relations on the North Queensland Frontier, 1861–1897*. Australian National University Press.

Ludwig, John A., Michael B. Coughenour, Adam C. Liedloff, and Rodd Dyer. 2001. "Modelling the Resilience of Australian Savanna Systems to Grazing Impacts." *Environment International* 27: 167–72.

MacDonald, D. H. 2018. *Before Yellowstone: Native American Archaeology in the National Park*. University of Washington Press.

Marder, M. 2013. *Plant-Thinking: A Philosophy of Vegetal Life*. Columbia University Press.

Margulies, Jared D. 2023. *The Cactus Hunters: Desire and Extinction in the Illicit Succulent Change*. University of Minnesota Press.

Marin, F. 2022. "Climate Change, Weather and Perception: Fishing in Eastern Patagonia." In *The Anthroposcene of Weather and Climate: Ethnographic Contributions to the Climate Change Debate*, edited by P. Sillitoe, 45–68. Berghahn Books.

Marshall, J. C., J. J. Blessing, S. E. Clifford, P. M. Negus, and A. L. Steward. 2020. "Epigeic Invertebrates of Pig-Damaged, Exposed Wetland Sediments Are Rooted: An Ecological Response to Feral Pigs (*Sus scrofa*)." *Aquatic Conservation: Marine and Freshwater Ecosystems* 30, no. 12: 2207–20. https://doi.org/10.1002/aqc.3468.

Martin, Aryn, Natasha Myers, and Ana Viseu. 2015. "The Politics of Care in Technoscience." *Social Studies of Science* 45, no. 5: 625–41. https://doi.org/10.1177/0306312715603249.

Martin, D. F. 1993. "Autonomy and Relatedness: An Ethnography of Wik People of Aurukun, Western Cape York Peninsula." PhD diss., Australian National University.

Martin, R. J., and D. Trigger. 2015. "Negotiating Belonging: Plants, People, and Indigeneity in Northern Australia." *Journal of the Royal Anthropological Institute* 21: 276–95.

May, D. 1994. *Aboriginal Labour and the Cattle Industry: Queensland from White Settlement to the Present.* Cambridge University Press.

McCarthy, Marty, Cassandra Hough, Hailey Renault, Carmen Brown, Charlie McKillop, Eliza Rogers, and Zara Margolis. 2014. "Biosecurity: Queensland Horticulture Poor Cousins to Livestock in Pest and Disease Funding." *ABC News*, November 28, 2014. www.abc.net.au/news/rural/2014-11-28/queensland-biosecurity-is-it-safe/5918944.

McClure, M. L., C. L. Burdett, M. L. Farnsworth, S. J. Sweeney, and R. S. Miller. 2018. "A Globally-Distributed Alien Invasive Species Poses Risks to United States Imperiled Species." *Scientific Reports* 8, no. 1: 1–9. https://doi.org/10.1038/s41598-018-23657-z.

McGrath, A. 1987. *Born in the Cattle: Aborigines in Cattle Country.* Allen and Unwin Australia.

Meat & Livestock Australia. 2007. *A Guide to Using NLIS Approved Ear Tags and Rumen Boluses.* Sydney, NSW.

Merlan, F. 1998. *Caging the Rainbow: Places, Politics, and Aborigines in a North Australian Town.* University of Hawai'i Press.

———. 2005. "Explorations towards Intercultural Accounts of Socio-Cultural Reproduction and Change." *Oceania* 75, no. 3: 167–82.

Meurk, C. 2011. "Loving Nature, Killing Nature, and the Crises of Caring: An Anthropological Investigation of Conflicts Affecting Feral Pig Management in Queensland, Australia." PhD diss., University of Queensland.

———. 2015. "Contesting Death: Conservation, Heritage, and Pig Killing in Far North Queensland, Australia." *Environmental Values* 24, no. 1: 79–104.

Mihailou, Helenna, and Melanie Massaro. 2021. "An Overview of the Impacts of Feral Cattle, Water Buffalo, and Pigs on the Savannas, Wetlands, and Biota of Northern Australia." *Austral Ecology* 46: 699–712. https://doi.org/10.1111/aec.13046.

Milne, Sarah. 2022. *Corporate Nature: An Insider's Ethnography of Global Conservation.* Critical Green Engagements: Investigating the Green Economy and Its Alternatives University of Arizona Press.

Moura, L. C., A. O. Scariot, I. B. Schmidt, R. Beatty, and Jeremy Russell-Smith. 2019. "The Legacy of Colonial Fire Management Policies on Traditional Livelihoods and Ecological Sustainability in Savannas: Impacts, Consequences, New Directions." *Journal of Environmental Management* 232: 600–606.

Mulcock, J., and D. Trigger. 2008. "Ecology and Identity: A Comparative Perspective on the Negotiation of 'Nativeness.'" In *Toxic Belonging? Identity and Ecology in Southern Africa*, edited by D. Wylie, 178–98. Cambridge Scholars Publishing.

Munn, N. 1970. "The Transformation of Subjects into Objects in Walbiri and Pitjantjatjara Myth." In *Australian Aboriginal Anthropology: Modern Stud-*

ies in the Social Anthropology of the Australian Aborigines, edited by R. Berndt, 141–62. University of Western Australia Press.

Myers, Fred R. 1986. *Pintupi Country, Pintupi Self: Sentiment, Place, and Politics among Western Desert Aborigines*. Smithsonian Institution Press.

———. 2002. "Ways of Place-Making." *La Ricerca Folklorica* 45: 101–19.

Nabhan, Gary P., Richard L. Knight, and Susan Charnely. 2014. "The Biodiversity That Protected Areas Can't Capture: How Private Ranch, Forest, and Tribal Lands Sustain Biodiversity." In *Stitching the West Back Together: Conservation of Working Landscapes*, edited by Susan Charnely, Thomas E. Sheridan, and Gary P. Nabhan, 33–47. University of Chicago Press.

Nabokov, P., and L. L. Loendorf. 2004. *Restoring a Presence: American Indians and Yellowstone National Park*. University of Oklahoma Press.

Nadasdy, P. 1999. "The Politics of TEK: Power and the 'Integration' of Knowledge." *Arctic Anthropology* 36, no. 1/2: 1–18.

———. 2003. *Hunters and Bureaucrats: Power, Knowledge and Aboriginal-State Relations in the Southwest Yukon*. University of British Columbia Press.

Nature Conservation Act, Queensland. 1992. Nature Conservation Act.

Neale, T. 2012. "'A Substantial Piece in Life': Viabilities, Realities, and Given Futures at the Wild Rivers Inquiries." *Australian Humanities Review* 53.

———. 2017. *Wild Articulations: Environmentalism and Indigeneity in Northern Australia*. University of Hawai'i Press.

———. 2019. "A Sea of Gamba: Making Environmental Harm Illegible in Northern Australia." *Science as Culture* 28, no. 4: 403–26. https://doi.org/10.1080/09505431.2018.1552933.

———. 2022. "Interscalar Maintenance: Configuring an Indigenous 'Premium Carbon Product' in Northern Australia (and Beyond)." *Journal of the Royal Anthropological Institute*. https://doi.org/10.1111/1467-9655.13861.

Neale, T., and J. M. Macdonald. 2019. "Permits to Burn: Weeds, Slow Violence, and the Extractive Future of Northern Australia." *Australian Geographer* 50, no. 4: 417–33. https://doi.org/10.1080/00049182.2019.1686731.

Negus, P. M., J. C. Marshall, S. E. Clifford, J. J. Blessing, and A. L. Steward. 2019. "No Sitting on the Fence: Protecting Wetlands from Feral Pig Damage by Exclusion Fences Requires Effective Fence Maintenance." *Wetlands Ecology and Management* 27: 581–85. https://doi.org/https://doi.org/10.1007/s11273-019-09670-7.

Neilly, Heather, and Lin Schwarzkopf. 2019. "The Impact of Cattle Grazing Regimes on Tropical Savanna Bird Assemblages." *Austral Ecology* 44: 187–98. https://doi.org/10.1111/aec.12663.

Norgaard, K. M. 2011. *Living in Denial: Climate Change, Emotions, and Everyday Life*. MIT Press.

NSW Threatened Species Scientific Committee. 2019. *1080 Poison Baiting Used for the Control of Vertebrate Pest Animals—Rejection of Key Threatening Process Listing*. NSW Department of Planning, Industry and Development. Sydney. www.environment.nsw.gov.au/topics/animals-and-plants/

threatened-species/nsw-threatened-species-scientific-committee/determina
tions/final-determinations/2008-2010/1080-poison-baiting-for-control-ver
tebrate-pest-animals-rejection-of-key-threatening-process-listing.

Nursey-Bray, M., H. Marsh, and H. Ross. 2010. "Exploring Discourses in Environmental Decision Making: An Indigenous Hunting Case Study." *Society
and Natural Resources* 23, no. 4: 366–82. https://doi.org/10.1080/08941920
903468621.

Ockwell, D., and Y. Rydin. 2006. "Conflicting Discourses of Knowledge: Understanding the Policy Adoption of Pro-Burning Knowledge Claims in Cape
York Peninsula, Australia." *Environmental Politics* 15, no. 3: 379–98.

Ogden, L. A. 2011. *Swamplife: People, Gators, and Mangroves Entangled in
the Everglades.* University of Minnesota Press.

Oliver-Smith, A. 1999. " 'What Is a Disaster?': Anthropological Perspectives on
a Persistent Question." In *The Angry Earth: Disaster in Anthropological
Perspective,* edited by A. Oliver-Smith and S. Hoffman, 18–34. Routledge.

Olwig, M. F., and L. V. Rasmussen. 2015. "West African Waterworlds: Narratives of Absence versus Narratives of Excess." In *Waterworlds: Anthropology in Fluid Environments,* edited by K. Hastrup and F. Hastrup, 110–28.
Berghahn Books.

Ottosson, Å. 2010. "Aboriginal Music and Passion: Interculturality and Difference in Australian Desert Towns." *Ethnos* 75, no. 3: 275–300.

———. 2012. "The Intercultural Crafting of Real Aboriginal Country and Man-
hood in Central Australia." *Australian Journal of Anthropology* 12: 179–
96.

———. 2016. " 'Don't Rubbish Our Town': 'Anti-Social Behaviour' and Indige-
nous-Settler Forms of Belonging in Alice Springs, Central Australia." *City
and Society* 28, no. 2: 152–73.

Pálsson, G. 1996. "Human-Environmental Relations: Orientalism, Paternalism,
and Communalism." In *Nature and Society: Anthropological Perspectives,*
edited by P. Descola and G. Pálsson, 63–81. Routledge.

Pascoe, B. 2014. *Dark Emu: Black Seeds Agriculture or Accident?* Magabala
Books.

Perry, J. J., M. Sinclair, H. Wikmunea, S. Wolmby, D. Martin, and B. Martin.
2018. "The Divergence of Traditional Aboriginal and Contemporary Fire
Management Practices on Traditional Lands, Cape York Peninsula, Northern Australia." *Ecological Management and Restoration* 19, no. 1: 24–31.

Petrie, Claire. 2019. *Live Export—a Chronology.* Parliament of Australia.
ACT, Canberra. www.aph.gov.au/About_Parliament/Parliamentary_Depart
ments/Parliamentary_Library/pubs/rp/rp1920/Chronologies/LiveExport.

Petty, A. 2013. "Field of Nightmares: Gamba Grass in the Top End." *The Conversation.*

Pierrehumbert, R. T., and G. Eshel. 2015. "Climate Impact of Beef: An Analysis
Considering Multiple Timescales and Production Methods with Use of
Global Warming Potentials." *Environmental Research Letters* 10 (085002).
https://doi.org/10.1088/1748-9326/10/8/085002.

Povinelli, E. 2011. "Routes/Worlds." *e-flux journal* 27. www.e-ux.com/journal/27/67991/routes-worlds/.

———. 2016. *Geontologies: A Requiem to Late Liberalism*. Duke University Press.

Puig de la Bellacasa, M. 2017. *Matters of Care: Speculative Ethics in More Than Human Worlds*. University of Minnesota Press.

Queensland Government. 2013. *Rinyirru (Lakefield) Aggregation Management Statement 2013*. Department of National Parks, Recreation, Sport and Racing. Brisbane. https://parks.des.qld.gov.au/__data/assets/pdf_file/0027/165771/rinyirru.pdf.

———. 2016. "Wild Dog Facts: The Law and Your Responsibility," edited by Department of Agriculture and Fisheries. State of Queensland.

———. 2017. "Pest Plants and Animals," edited by Department of Environment and Science. State of Queensland.

———. 2022. "National Livestock Identification System." State of Queensland. Accessed September 7, 2023. www.business.qld.gov.au/industries/farms-fishing-forestry/agriculture/animal/move/laws/nlis.

Queensland Land Tribunal. 1996. *Aboriginal Land Claim to Lakefield National Park/Report of the Land Tribunal Established under the Aboriginal Land Act 1991 to the Hon the Minister for Natural Resources*. The Tribunal Brisbane.

Qviström, M. 2007. "Landscapes Out of Order: Studying the Inner Urban Fringe beyond the Rural-Urban Divide." *Geografiska Annaler: Series B, Human Geography* 89, no. 3: 269–82. https://doi.org/10.1111/j.1468-0467.2007.00253.x.

Radomski, P., and D. Perleberg. 2019. "Avoiding the Invasive Trap: Policies for Aquatic Non-Indigenous Plant Management." *Environmental Values* 28, no. 2: 211–32. https://doi.org/10.3197/096327119X15515267418539.

Reardon-Smith, M. 2021. "'We Are the Ones Who Know the Intimacies of the Soil': Grazier Claims to Belonging and Changing Land Relations in Cape York Peninsula, Queensland." *Australian Journal of Anthropology* 32, no. 3: 340–56.

———. 2023a. "Grappling with Weeds: Invasive Species and Hybrid Landscapes in Cape York Peninsula, Far Northeast Australia." *Environmental Values* 32, no. 3: 249–69. https://doi.org/10.3197/096327122X16491521047044.

———. 2023b. "Valuing Hard Work: 'Station Times,' the Pioneer Complex and Settler-Descended Graziers' Views on Work in Cape York Peninsula." *Asia Pacific Journal of Anthropology*: 1–19. https://doi.org/10.1080/14442213.2023.2234336.

———. 2024. "A Seat at the Table? Planning, Meetings and the 'Stable Relation' of a Joint Managed National Park in Northern Australia." *Geoforum* 150. https://doi.org/10.1016/j.geoforum.2024.103992.

———. 2025. "'It's about Getting the Right People Back on the Right Country!': Cultural Difference and Structural Inequality in a Northern Australian Joint Managed National Park." *Journal of Political Ecology*.

Recio, E., and D. Hestad. 2022. *Indigenous Peoples: Defending an Environment for All*. International Institute for Sustainable Development. www.iisd .org/system/files/2022-04/still-one-earth-Indigenous-Peoples.pdf.

Redmond, A. 2005. "Strange Relatives: Mutualities and Dependencies between Aborigines and Pastoralists in the Northern Kimberley." *Oceania* 75, no. 3: 234–46.

Richardson, Krystle. 2017. "Why Is the Government Killing Perfectly Good Cattle?" *Machines4U: Farming News*, April 18, 2017. Accessed June 23, 2023. www.machines4u.com.au/mag/why-is-the-government-killing-perfect ly-good-cattle/.

Rigsby, B. 1981. "Aboriginal People, Land Rights, and Wilderness on Cape York Peninsula: Presidential Address Delivered before the Royal Society of Queensland (31 March 1980)." *Proceedings of the Royal Society of Queensland* 92: 1–10.

Riley, Sophie. 2013. " 'Buffalo belong Here, as Long as He Doesn't Do too Much Damage': Indigenous Perspectives on the Place of Alien Species in Australia." *Australasian Journal of Natural Resources Law and Policy* 16, no. 2: 157–96.

———. 2019. "Horses, Culture, and Ethics: Wildlife Regulation in Kosciusko National Park." *Environmental Planning and Law Journal* 36: 674–91.

Ritchie, D. 2009. "Things Fall Apart: The End of an Era of Systematic Indigenous Fire Management." In *Culture, Ecology, and Economy of Fire Management in North Australian Savannas: Rekindling the Wurrk Tradition*, edited by Jeremy Russell-Smith, P. Whitehead, and P. Cooke, 23–40. CSIRO Publishing.

Robbins, P., and S. A. Moore. 2012. "Ecological Anxiety Disorder: Diagnosing the Politics of the Anthropocene." *Cultural Geographies* 20, no. 1: 3–19. https://doi.org/10.1177/1474474012469887.

Robbins, Paul. 2004. "Comparing Invasive Networks: Cultural and Political Biographies of Invasive Species." *Geographical Review* 94, no. 2: 139–56. https://doi.org/10.1111/j.1931-0846.2004.tb00164.x.

———. 2007. *Lawn People: How Grasses, Weeds, and Chemicals Make Us Who We Are*. Temple University Press.

Roberts, J. 2019. " 'We Live Like This': Local Inequalities and Disproportionate Risk in the Context of Extractive Development and Climate Change on New Hanover Island, Papua New Guinea." *Oceania* 89, no. 1: 68–88.

Robinson, C. J., D. Smyth, and P. J. Whitehead. 2005. "Bush Tucker, Bush Pets, and Bush Threats: Cooperative Management of Feral Animals in Australia's Kakadu National Park." *Conservation Biology* 19, no. 5: 1385–91.

Rodríguez, I., B. Sletto, B. A. Bilbao, I. Sánchez-Rose, and A. Leal. 2018. "Speaking of Fire: Reflexive Governance in Landscape of Social Change and Shifting Local Identities." *Journal of Environmental Policy and Planning* 20, no. 6: 689–703. https://doi.org/10.1080/1523908X.2013.766579.

Rose, D. B. 1992. *Dingo Makes Us Human: Life and Land in an Australian Aboriginal Culture*. Cambridge University Press.

———. 2004. *Reports from a Wild Country: Ethics for Decolonisation*. UNSW Press.

Rowland, M. J. 2004. "Return of the 'Noble Savage': Misrepresenting the Past, Present, and Future." *Australian Aboriginal Studies* 2: 2–14.

Russell-Smith, Jeremy, G. D. Cook, P. M. Cooke, A. C. Edwards, M. Lendrum, C. P. Meyer, and P. Whitehead. 2013. "Managing Fire Regimes in Northern Australian Savannas: Applying Aboriginal Approaches to Contemporary Global Problems." *Frontiers in Ecology and the Environment* 11, Special Issue 1: 55–63.

Sackett, Lee. 1991. "Promoting Primitivism: Conservationist Depictions of Aboriginal Australia." *Australian Journal of Anthropology* 2, no. 2: 233–46.

Sakadevan, K., and M.-L. Nguyen. 2017. "Livestock Production and Its Impact on Nutrient Pollution and Greenhouse Gas Emissions." *Advances in Agronomy* 141: 147–84. https://doi.org/https://doi.org/10.1016/bs.agron.2016.10.002.

Sayre, Nathan F. 2007. "A History of Working Landscapes: The Altar Valley, Arizona, USA: How Ranchers Have Shaped the West—and Continue to Do So." *Rangelands* 29, no. 3: 41–45.

Scaramelli, Caterina. 2021. *How to Make a Wetland: Water and Moral Ecology in Turkey*. Stanford University Press.

Schmidt, I. B., and L. Eloy. 2020. "Fire Regime in the Brazilian Savanna: Recent Changes, Policy, and Management." *Flora* 268: 151–63.

Secretariat of the Convention on Biological Diversity. 2020. *Global Biodiversity Outlook 5—Summary for Policy Makers*. Montreal. www.cbd.int/gbo/gbo5/publication/gbo-5-spm-en.pdf.

Seton, K. A., and J. Bradley. 2004. "'When You Have No Law You Are Nothing': Cane Toads, Social Consequences, and Management Issues." *Asia Pacific Journal of Anthropology* 5, no. 3: 205–25.

Shaffer, L. J. 2010. "Indigenous Fire Use to Manage Savanna Landscapes in Southern Mozambique." *Fire Ecology* 6: 43–59.

Sheridan, Thomas E., Nathan F. Sayre, and David Seibert. 2014. "Beyond 'Stakeholders' and the Zero-Sum Game: Towards Community-Based Conservation in the American West." In *Stitching the West Back Together: Conservation of Working Landscapes*, edited by Susan Charnely, Thomas E. Sheridan, and Gary P. Nabhan, 53–80. University of Chicago Press.

Sherley, M. 2007. "Is Sodium Fluoroacetate (1080) a Humane Poison?" *Animal Welfare* 16: 449–58.

Shotwell, A. 2016. *Against Purity: Living Ethically in Compromised Times*. University of Minnesota Press.

Simone, T. 2016. "Aboriginal Stockwomen: Their Legacy in the Australian Pastoral Industry." PhD diss., Deakin University.

Singh, B., and Naisargi N. Davé. 2015. "On the Killing and Killability of Animals: Nonmoral Thoughts for the Anthropology of Ethics." *Comparative Studies of South Africa and the Middle East* 35, no. 2: 232–45.

Slater, L. 2013. "'Wild Rivers, Wild Ideas': Emerging Political Ecologies of

Cape York Wild Rivers." *Environment and Planning D: Society and Space* 31: 763–78. https://doi.org/10.1068/d3012.

Sletto, B. 2008. "The Knowledge That Counts: Institutional Identities, Policy Science, and the Conflict over Fire Management in the Gran Sabana, Venezuela." *World Development* 36, no. 10: 1938–55.

———. 2011. "Conservation Planning, Boundary-Making, and Border Terrains: The Desire for Forest and Order in the Gran Sabana, Venezuela." *Geoforum* 42: 197–210.

Smee, B. 2019. "Up to 500,000 Drought-Stressed Cattle Killed in Queensland Floods." *The Guardian*, February 11, 2019. Accessed January 14, 2020. www.theguardian.com/australia-news/2019/feb/11/up-to-500000-drought -stressed-cattle-killed-in-queensland-floods.

Smith, B. R. 2003a. "Pastoralism, Local Knowledge, and Australian Aboriginal Development in Northern Queensland." *Asia Pacific Journal of Anthropology* 4, no. 1–2: 88–104.

———. 2003b. "Whither 'Certainty'? Coexistence, Change, and Land Rights in Northern Queensland." *Anthropological Forum* 13, no. 1: 27–48.

Stensrud, A. B. 2016. "Climate Change, Water Practices, and Relational Worlds in the Andes." *Ethnos* 81, no. 1: 75–98.

Stephan, K. 2006. *Cape York Peninsula Feral Pig Management Plan, 2006–2009.* Cape York Weeds and Feral Animals Program. Cooktown. https://capeyorknrm.com.au/sites/default/files/2019-04/cynrm025_pest-mgt_cywfap_feral-pig-mgt-plan.pdf.

Strang, V. 1997. *Uncommon Ground: Cultural Landscapes and Environmental Values.* Berg.

———. 2004a. "Knowing Me, Knowing You: Aboriginal and European Concepts of Nature as Self and Other." *Worldviews* 9, no. 1: 25–56.

———. 2004b. *The Meaning of Water.* Berg.

———. 2013. "Conceptual Relations: Water, Ideologies, and Theoretical Subversions." In *Thinking with Water,* edited by C. Chen, J. MacLeod, and A. Neimanis, 185–211. McGill-Queen's University Press.

———. 2014. "Fluid Consistencies: Material Relationality in Human Engagements with Water." *Archaeological Dialogues* 21, no. 2: 133–50.

Sutton, P. J. 1978. "Wik: Aboriginal Society, Territory, and Language at Cape Keerweer, Cape York Peninsula, Australia." PhD diss., Department of Anthropology and Sociology, University of Queensland.

The Guardian. 2011. "Australia Suspends Cattle Export to Indonesian Abattoirs." *The Guardian,* May 31, 2011. www.theguardian.com/world/2011/may/31/australia-suspends-cattle-export-indonesia.

Toussaint, Sandy. 2008. "Kimberley Friction: Complex Attachments of Water-Places in Northern Australia." *Oceania* 78: 46–61.

Toussaint, Sandy, Patrick Sullivan, Sarah Yu, and Mervyn Jnr. Mularty. 2001. *Fitzroy Valley Indigenous Cultural Values Study (a Preliminary Assessment).* Centre for Anthropological Research, University of Western Australia, Nedlands.

Trigger, D. 2008. "Indigeneity, Ferality, and What 'Belongs' in the Australian Bush: Aboriginal Responses to 'Introduced' Animals and Plants in a Settler-Descendant Society." *Journal of the Royal Anthropological Institute* 14, no. 3: 628–46.

———. 2012. "Whales, Whitefellas, and the Ambiguity of 'Nativeness': Reflections on the Emplacement of Australian Identities." In *Invasive and Introduced Plants and Animals: Human Perceptions, Attitudes, and Approaches to Management*, edited by I. D. Rotherham and R. A. Lambert, 109–20. Earthscan.

———. 2013. "Rethinking Nature and Nativeness." In *Up Close and Personal: On Peripheral Perspectives and the Production of Anthropological Knowledge*, edited by C. Shore and S. Trnka, 142–57. Berghahn Books.

Trigger, D., Y. Toussaint, and J. Mulcock. 2010. "Ecological Restoration in Australia: Environmental Discourses, Landscape Ideals, and the Significance of Human Agency." *Society and Natural Resources* 23, no. 11: 1060–74. https://doi.org/10.1080/08941920903232902.

Tsing, A. 2005. *Friction: An Ethnography of Global Connection*. Princeton University Press.

———. 2010. "Arts of Inclusion, or How to Love a Mushroom." *Mānoa* 22, no. 2: 191–203.

———. 2015. *The Mushroom at the End of the World: On the Possibility of Life in Capitalist Ruins*. Princeton University Press.

Tsing, A., A. S. Mathews, and N. Bubandt. 2019. "Patchy Anthropocene: Landscape Structure, Multispecies History, and the Retooling of Anthropology." *Current Anthropology* 60 (S20): S186–S197.

Turton, S. M. 2008. "Landscape-Scale Impacts of Cyclone Larry on the Forests of Northeast Australia, Including Comparisons with Previous Cyclones Impacting the Region between 1858 and 2006." *Austral Ecology* 33: 409–16.

Tweddle, Neil E., and Paul Livingstone. 1994. "Bovine Tuberculosis Control and Eradication Programs in Australia and New Zealand." *Veterinary Microbiology* 40: 23–39.

Vaarzon-Moral, P. 2017. "Alien Relations: Ecological and Ontological Dilemmas Posed for Indigenous Australians in the Management of 'Feral' Camels on Their Lands." In *Entangled Territorialities: Negotiating Indigenous Lands in Australia and Canada*, edited by F. Dussart and S. Poirier, 186–211. University of Toronto Press.

Valderama, S. P. 2020. "Disappearing Waste and Wasting Time: From Productive Fallows to Carbon Offset Production in Madagascar's Forests." *Ethnos*. https://doi.org/10.1080/00141844.2020.1796737.

van Dooren, Thom. 2011. "Invasive Species in Penguin Worlds: An Ethnical Taxonomy of Killing for Conservation." *Conservation and Society* 9, no. 4: 286–98.

———. 2014. *Flight Ways: Life and Loss at the Edge of Extinction*. Columbia University Press.

———. 2019. *The Wake of Crows: Living and Dying in Shared Worlds*. Critical

Perspectives on Animals: Theory, Culture, Science, and Law. Columbia University Press.

van Dooren, Thom, E. Kirksey, and Ursula Münster. 2016. "Multispecies Studies: Cultivating Arts of Attentiveness." *Environmental Humanities* 8, no. 1: 1–23. https://doi.org/10.1215/22011919-3527695.

van Voorst, Roanne. 2014. "The Right to Aid: Perceptions and Practices of Justice in a Flood-Hazard Context in Jakarta, Indonesia." *Asia Pacific Journal of Anthropology* 15, no. 4: 339–56.

Von Sturmer, J. R. 1978. "The Wik Region: Economy, Territoriality, and Totemism in Western Cape York Peninsula, North Queensland." PhD diss., Department of Anthropology and Sociology, University of Queensland.

Wadiwel, Dinesh. 2023. *Animals and Capital.* Edinburgh University Press.

Wanderer, Emily. 2020. *The Life of a Pest: An Ethnography of Biological Invasion in Mexico.* University of California Press.

Watt, Laura Alice. 2017. *The Paradox of Preservation: Wilderness and Working Landscapes at Point Reyes National Seashore.* University of California Press.

Weeks, Kathi. 2011. *The Problem with Work: Feminism, Marxism, Antiwork Politics, and Postwork Imaginaries.* Duke University Press.

West, P. 2006. *Conservation Is Our Government Now: The Politics of Ecology in Papua New Guinea.* Duke University Press.

West, P., J. Igoe, and D. Brockington. 2006. "Parks and Peoples: The Social Impact of Protected Areas." *Annual Review of Anthropology* 35: 251–77.

Yibarbuk, D., P. J. Whitehead, Jeremy Russell-Smith, D. Jackson, C. Godjuwa, A. Fisher, P. Cooke, D. Choquenet, and D. M. J. S. Bowman. 2001. "Fire Ecology and Aboriginal Land Management in Central Arnhem Land, Northern Australia: A Tradition of Ecosystem Management." *Journal of Biogeography* 28, no. 3: 325–43.

Yocum, H. M. 2016. "'It *Becomes* Scientific . . .': Carbon Accounting for REDD+ in Malawi." *Human Ecology* 44: 677–85. https://doi.org/10.1007/s10745-016-9869-y.

INDEX

Davé, Naisargi, 13, 16, 126, 207,
220. *See also* Singh, Bhrigupati
and Naisargi Davé
dingoes, 124, 133
disaster relief, 191–193
disasters, 192
dispossession, 20, 51–52, 65, 111
drought, 181, 187
dry season, 16–17

endangered species, 129, 211–212
environmental knowledges, 4
environmental protection legislation,
3, 109, 133
ethics of care. *See* care
Equal Wages legislation, 17, 57, 64;
graziers and, 58, 85, 90
erosion, 88–89, 117; coastal, 194–195

fences, 53
feral animals, 30–31, 82, 118; cattle,
72, 76–77, 82; as a food source,
58, 131. *See also* cattle; feral dynamics; feral pigs
feral pigs, 31, 132, 136; cost, 124–
125; damage to cultural sites, 134;
environmental damage, 122, 126,
128, 130, 133; as food source, 58,
131
feral dynamics, 110, 118–119
fire: aerial burning, 155, 158–159;
carbon credits scheme, 163–164,
166; management, 160, 177;
mosaic burns, 155, 166, 172. *See
also* cool burns; cultural burning;
fire regimes; storm burns; wildfires
firefighting, 173–174
fire regimes, 32, 156, 162; graziers',
172; Indigenous, 151, 225n2; "natural," 169–172, 178; Queensland
Parks', 160–162; viability, 107. *See
also* cool burns, cultural burning;
fire, fire fighting; storm burns
flooding, 186, 189–190; cattle and,
191. *See also* wet season

forest fires. *See* wildfires
frontier violence, 51

Galvin, Shaila Seshia, 15, 117–118
gamba grass, 101, 110–111, 118; utility, 117
gold rush, 51
grader grass, 101
grasses, 101; introduced 109–111,
117; native, 73, 118
grazing. *See* cattle grazing
graziers, 39, 61, 92; and Aboriginal
people, 58; bush fires, 174–175;
climate change views, 199–200,
202–208; conservation and, 91;
controlled burning, 156–157, 165;
livelihoods, 38, 46, 49, 83, 85; relationship with land, 92, 95, 117,
171; relationship with park authorities, 70, 79–81, 84, 87, 95.
See also cattle grazing; live export
Great Barrier Reef, 18; sediment runoff, 88–89

Haraway, Donna, 8, 32, 43–44, 65,
126, 212
Hastrup, Kirsten, 187, 198–199
Head, Lesley, 102–103, 108, 118,
132–133
herbicide, 100, 103–104. *See also*
weed control
horses, 53, 224n2; culling, 78, 82
hurricanes. *See* cyclones

illegal hunting, 31, 142; in Florida,
80
Indigenous people, 17; in American
Mid-west, 17; of Hawai'i, 132;
Inuit, 196; Q'eqchi' people, 52. *See
also* Aboriginal Traditional
Owners
Intergovernmental Panel on Climate
Change (IPCC), 198
introduced species, 4; Aboriginal Tra-
ditional Owners and, 136; fruit

The authorized representative in the EU for product safety and compliance is:
Mare Nostrum Group
B.V Doelen 72
4831 GR Breda
The Netherlands

www.ingramcontent.com/pod-product-compliance
Lightning Source LLC
Chambersburg PA
CBHW020848270326
41928CB00006B/598